THE MYSTIC BEAUTY OF THE TREFOIL

James R Warren

BLOXWICH
2026

Published in the United Kingdom in 2026 by
Midland Tutorial Productions

First Edition 1 February 2026

Cover Design: James Warren

File Prefix Code: MBT

ISBN 978 1 915750 16 7

Email: MTP1586@gmail.com

COVER PHOTOGRAPH:
Lichfield Cathedral from Stowe Pool
31 March 2009

THE MYSTIC BEAUTY OF THE TREFOIL

James R Warren

MIDLAND TUTORIAL

MIDLAND TUTORIAL PRODUCTIONS
BLOXWICH

Other Books By James R Warren

Boscawen-Ûn
Beyond Tourist Britain
Gleanings as I Pass
Exordium
Meditations
Gamma Solution
Moddeshall Hydropower
Unreasonable Mathematics
Mathematical Explorations
Researches: Volume One
Researches: Volume Two
Researches: Volume Three
Researches: Volume Four
Pi and Phi
Four Famous Numbers
Progress in Iron Smelting
Congleholme
Acts of Trust
TAM

To The Glory of
The Loving God

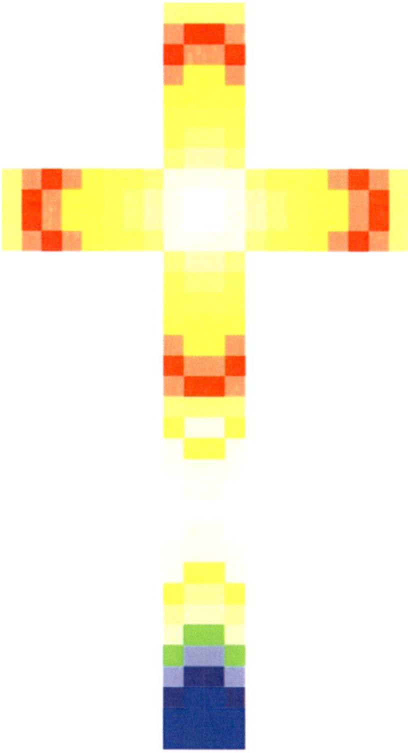

Who Made Our Minds Free

MBT

WARREN

TABLE OF CONTENTS

Contents

MBT

WARREN

PROLOG

In the beginning was the Word, and the Word was with God, and the Word was God.

The same was in the beginning with God.

All things were made by him; and without him was not any thing made that was made.

In him was life; and the life was the light of men.

And the light shineth in darkness; and the darkness comprehended it not.

(John 1:1-5)

What is The Word? Is a word a coded plan or is it something else: *anything*?

Certainly, after years deliberating the trackless aether we have decided that *something* inhabits *nothing*; or that the nothing can be an infinitely attenuated gulph of matter, not supportive of light; or perhaps that nothing is impossible, *literally*.

But what are years and what is light?

Does the vast emptiness of space have to bear light to be perceptible to men?

As I remark elsewhere, when I was a small boy my Late Father taught me that matter is made of atoms and that almost all of an atom is void. Before that occasion, Father had bought me a great book called "Odham's Book of Knowledge for Boys and Girls". I was three. The book taught me to read. I avidly consumed it, every picture every word until its covers fell apart and disappeared, leaving me with its content, *whatever* that was, marks of carbon on reconstituted cellulose, or *something* else, *anything* else.

I accepted that it was what it said on the title page and the long-gone cover, a book of knowledge. This is what my Father had always called it. Perhaps a Tree of Knowledge appeared somewhere near the front, whilst a fully-clothed boy and girl aspired to pick its fruit. The fruit was depicted somewhat literally as nine gaily-colored squares of the letters K, N, O, W, L, E, D, G and E. Sadly, I am one of the last of the Old British. In my universe the literal is eminently capable of being *somewhat*.

I was an old man before I questioned the *spirit* of the book, *any* book, or its ontology, or the ontology of *anything*, or even knew what ontology was, assuming I now know, for as we have seen knowledge is impossible. At one level my Father was a bag of blood and bones, at another just an ordinary seaman, and at other levels *something else entirely*.

As a seventy-three-year-old I retrieved my sorry copy of the Odham's tome and after considerable research provisionally identified this work as "Odham's Encyclopaedia for Children", published in 1955 by Odham's of Long Acre, London, and edited by JA Lauwerys, RW Moore and Brian Vesey-Fitzgerald.

I often wonder what it was exactly that the Greeks understood by λογος: was it Knowledge, such as a God would have needed to fashion his Product, and if so whence came that Knowledge, in a primal void without books and without teachers?

Clearly there is something immature about the question, but deeper perplexities remain about the nature of Creation.

For sure, we have adopted the term λογος or latinized it to denote a science, but this merely atomises and evades the greater question: What is Knowledge, and even is it possible?

If light is ubiquitous in space is it really so, or inherent only in those parts men and their machines can see?

Or as some philosophers say, does light not emerge to existence until witnessed, a wholly anthropogenic phenomenon? If so then is light like the light of the burglar's Hand of Glory, carrying the candle whose light lights his way without waking his victim from his slumbers, or alerting the watchman? Such ideas seem irrational and solipsistic, but that does not mean untenable, at least not for something as counterintuitive as electromagnetic radiation.

As for λογος is it merely a dismissive profanity, shouted in the street, a lazy word to cover a multitude of sins to describe the indescribable, or is it *something else entirely*?

And was the Word coeval with God, *whenever* he and it arose, or was it "there", so to say, waiting for God to come and reify it?

For sure, the Trefoil is a very simple object, if an object it is. It is one of the few geometric figures that is constructible using *a pair of compasses only*. If the Greeks did not know what a λογος was then they knew how to make a Trefoil and masons knew how to fashion one.

As I age I become ever more convinced that we should Remember many and much else. Let us then remember the splendid Alfred Joyce Kilmer (1886-1918) who like so many died before his time at The Second Battle of the Marne:-

Trees

I think that I shall never see

A poem lovely as a tree.

A tree whose hungry mouth is prest
Against the earth's sweet flowing breast;

A tree that looks at God all day,
And lifts her leafy arms to pray;

A tree that may in summer wear
A nest of robins in her hair;

Upon whose bosom snow has lain;
Who intimately lives with rain.

Poems are made by fools like me,
But only God can make a tree.

It is a simple poem and much-mocked but like all simple things accommodates a deeper Truth. I am a worse poet than Kilmer and a much worse man, but as I did before a third Man who once addressed me with a simple and much-mocked Message I stand back and wonder.

The Second Battle of the Marne was a very big deal to the men of its time and indeed it transfigured the trajectory of world history. Today, we have, beside the powers to devastate Picardy or any province, the powers to annihilate a planet, or maybe several of those.

To survive we must literally think the unthinkable. That is a paradox and a puzzle, but the demons of the poets of Classical Antiquity were one and all notorious for their baffling riddles, solved by heroes.

CHAPTER ONE
THE ANATOMY OF A TREFOIL

A trefoil is a simple planar figure of three lobes. The trefoils we shall discuss are constructed of three circles which share a common Radius, R, which for convenience we shall define as unity, i.e. $R = 1$.

Trefoils are useful for illustrating the qualitative associations of three attributes.

This most basic radius-is-unity trefoil we will call a Blob, and it is illustrated below:-

Figure 1.1
BLOB
A Simple Trefoil Outline for
Three Intersecting Circles of Unitary Radii

The trefoil is one of the few planar figures you can construct using *a pair of compasses only*: No straight-edge is needed.

The trefoil is very easy to construct.

First draw a full circle and then without altering the set of your compasses draw a second circle from the circumference of the first like so:-

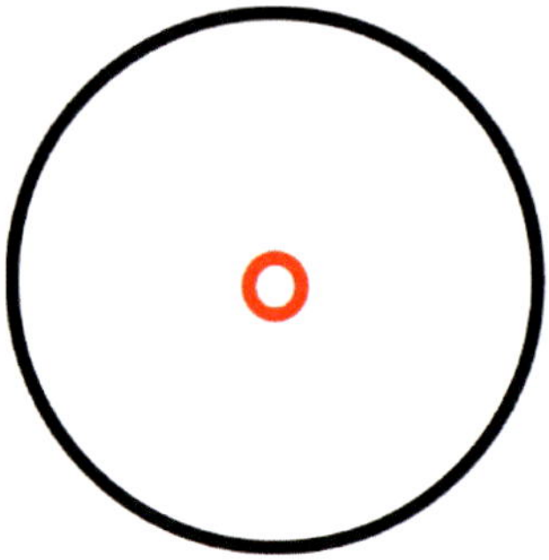

Draw First Circle

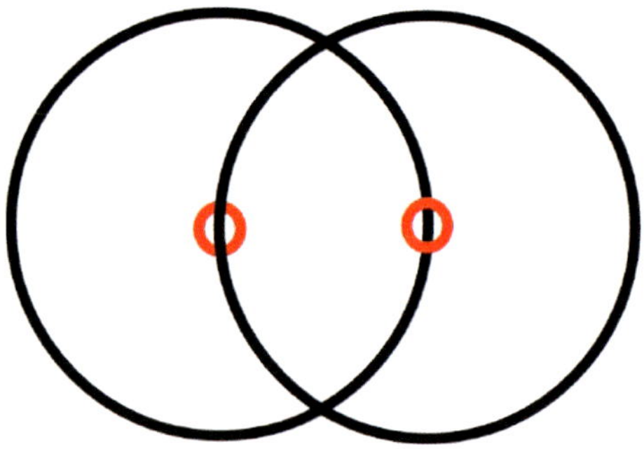

Add Second Circle

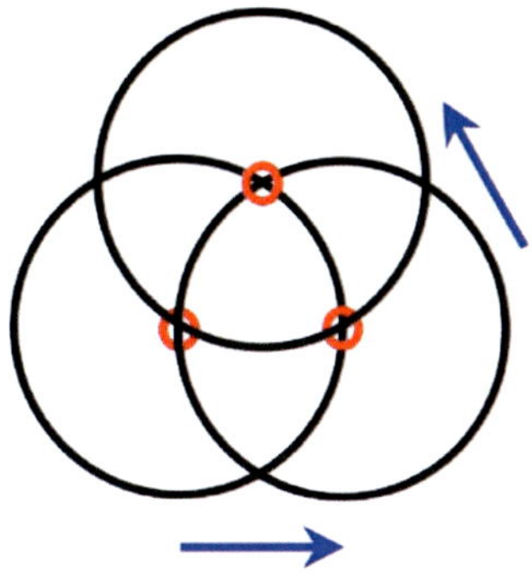

Add Third Circle

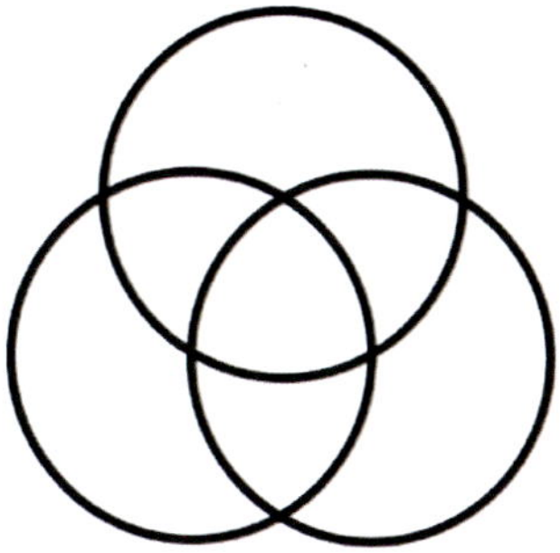

The Completed Figure

Figure 1.2
Stages in the Drafting of a Trefoil

The centers of the small red circles indicate where next to place the point of your compass.

As you can see the resulting trefoil has seven distinct zones whose vertices are circles' intersections. There is what appears to be, and indeed can be shown to be, a Reuleaux

Triangle (RT) at the center of the trefoil. This is bounded by three Circular Triangles (CT) which was will call Linguoids for short. Touching Pairs of Circular Triangles are three peripheral Lobes.

Figure 1.3 clarifies these sub-shape species:-

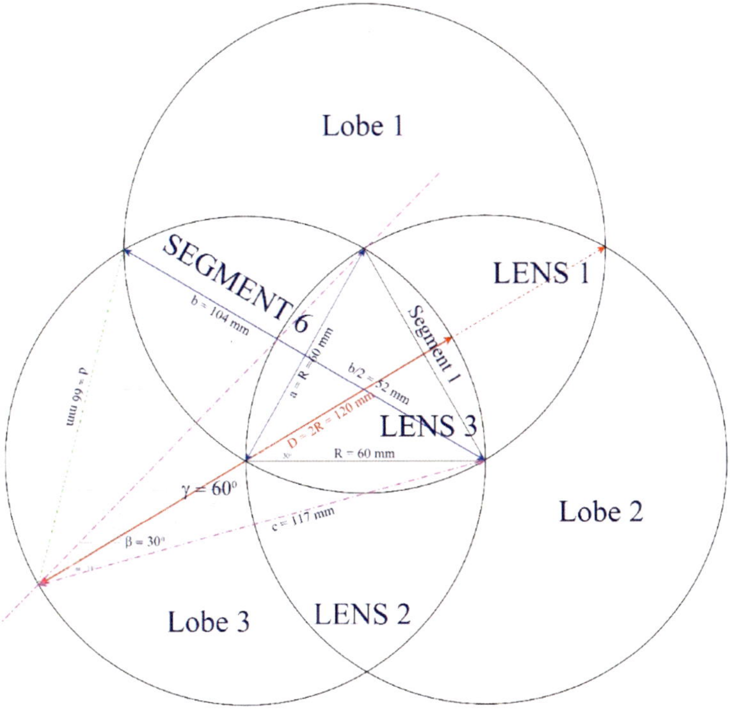

Figure 1.3
The Anatomy of a Trefoil

Three small Segments: Segment 1, Segment 2 and Segment 3 bound the central Equilateral Triangle (ET) to form the Reuleaux Triangle: And six larger SEGMENTS: SEGMENT 1 through SEGMENT 6 can be paired to form three Lenses which constitute the three overlapping areas between the component circles.

The Central Equilateral Triangle (ET)

The three sides of ET are each R in length and this length defines the size of the total trefoil figure.

For convenience and generality, R is unity.

Hence the Altitude b/2 is given by:-

$$b/2 = R\sin(60°) = 0.866025403784439$$

Equation 1.1

The Major Lens Diameter, b, is double this value and therefore:-

$$b = 2R\sin(60°) = 1.73205080756888$$

Equation 1.2

The Area of the Central Equilateral Triangle, AET, is thus:-

$$A_{ET} = \frac{R}{2} \times \frac{b}{2} = R^2 . \frac{\sqrt{3}}{4} = 0.433012701892219$$

Equation 1.3

The Isosceles Triangle Side, c

By the Sine Rule:-

$$c = \frac{\sin(90°)}{\sin(\alpha)} \cdot \frac{R}{2} = 1.93185165257814$$

Equation 1.4

Hence:-

$$c = \frac{4}{\sqrt{6}-\sqrt{2}} \cdot \frac{R}{2} = \frac{2R}{\sqrt{6}-\sqrt{2}} = 1.93185165257814$$

Equation 1.5

The Chord and Altitude of the Minor Segment 1

As you can see from the diagram the Chord, a, is equivalent to unity and is defined by:-

$$a = 2c\sin(15°) = 1$$

Equation 1.6

D₋ is the Altitude of the constructional isosceles triangle whose long sides are c and is given by:-

$$D_{-} = c.\cos(15°) = c\frac{\sqrt{6}+\sqrt{2}}{4} = 1 + \frac{\sqrt{3}}{2}$$

Equation 1.7

whilst D is twice the Circle Radius, R. Hence the Minor Segment Altitude, MSA, is given by:-

$$MSA = D - D_{-} = 2R - \left(1 + \frac{\sqrt{3}}{2}\right) = 0.133974596215561$$

Equation 1.8

The Chord and Altitude of the Major SEGMENT 6

By definition, the Altitude of a Major SEGMENT, NSA, is one half of Central Equilateral Side Length, a, or:-

$$NSA = \frac{a}{2} = \frac{1}{2}$$

Equation 1.9

given that R is unity.
The Chord of a Major SEGMENT, b, is clearly:-

$$b = 2\frac{b}{2}$$

Equation 1.10

The Area of Major SEGMENT 6

The Area of the Sector containing SEGMENT 6, A_{sect}, is given by:-

$$A_{sect} = \frac{1}{3} \cdot \pi \cdot R^2 = 1.0471975511966$$

Equation 1.11

whilst the Area of the Triangle subtending the SEGMENT 6 is:-

$$A_{tri} = \frac{1}{2} \cdot R^2 \cdot \sin(\delta) = \frac{1}{2} \cdot R^2 \cdot \sin(\delta)$$
$$= 0.433012701892219$$

Equation 1.12

Hence:-

$$A_{SEG} = A_{sect} - A_{tri} = 0.614184849304378$$

Equation 1.13

Alternatively, as the difference between the Area of the Containing Circle, A_c, and the Area of a Regular n-Gon, A_p we may write:-

$$A_{SEG} = A_c - A_p = \pi R^2 - \frac{1}{4} \cdot n \cdot s^2 \cdot \cot\left(\frac{\pi}{n}\right)$$

Equation 1.14

where n = 3 and s ≡ b.

For the case of R = 1, A_c is 3.14159265358979 and A_p is 1.29903810567666.

Accordingly:-

$$A_{SEG} = \frac{A_c - A_p}{n} = 0.614184849304378$$

Equation 1.15

Note also that:-

$$A_p = \frac{1}{4} \cdot n \cdot s^2 \cdot \cot\left(\frac{\pi}{n}\right) = \frac{1}{2} . b^2 \cdot \frac{\sqrt{3}}{2}$$

Equation 1.16

The Area of Minor Segment 1

For the minor segments the operative angle γ is 60° so that 3γ is π.

Correspondingly, the Area of the Minor Segment, A_{seg}, is given by:-

$$A_{seg} = A_{sect} - A_{tri} = \frac{1}{n} \cdot \pi R^2 - \frac{1}{2} \cdot R^2 \cdot \sin(\gamma)$$

Equation 1.17

or:-

$$A_{seg} = A_{sect} - A_{tri} = \frac{R^2}{2}\left[\frac{1}{3} \cdot \pi - \sin(\gamma)\right] = 0.09058607370608$$

Equation 1.18

where n = 6 and s ≡ R ≡1.

Alternatively, as the difference between the Area of the Containing Circle, A_c, and the Area of a Regular n-Gon, A_p we may write:-

$$A_{seg} = \frac{A_c - A_p}{n} = \frac{\pi R^2 - \frac{1}{4}\cdot n \cdot s^2 \cdot \cot\left(\frac{\pi}{n}\right)}{n}$$
$$= 0.09058607370608$$

Equation 1.19

The Area of the Central Reuleaux Triangle, A_{RT}

The Central Reuleaux Triangle is designated RT and its Area A_{RT}. Meanwhile, the Area of the Central Equilateral Triangle called ET is A_{ET}.

$$A_{RT} = A_{ET} + 3A_{seg} = 0.704770923010458$$

Equation 1.20

Alternatively, by the Blaschke-Lebesgue Theorem:-

$$A_{RT} = \frac{R^2}{2}\left(\pi - \sqrt{3}\right) = 0.704770923010458$$

Equation 1.21

Area of LENS3

The Area of a Lens, AL, is:-

$$A_L = 2A_{SEG} = 1.22836969860876$$

Equation 1.22

Area of a Circular Triangle CT ("Linguoid")

A Linguoid is designated CT and its Area is A_{CT}.

$$A_{CT} = \frac{A_{ET} + 2A_{seg}}{2} - A_{seg} = 0.523598775598299$$

Equation 1.23

Area of a Lobe

$$A_{Lobe} = \pi R^2 - (2A_{CT} + A_{RT}) = \frac{R^2}{6}\left(\pi - 3\sqrt{3}\right)$$

$$= 1.38962417938274$$

Equation 1.24

Area of the Blob

$$A_{Blob} = A_{RT} + 3(A_{Lobe} + A_{CT}) = \frac{R^2}{2}\left(3\pi - 2\sqrt{3}\right)$$

$$= 6.44443978795357$$

Equation 1.25

Equations Summary Diagram

Figure 1.4 summaries the Component Area Equations over their respective Trefoil Components:-

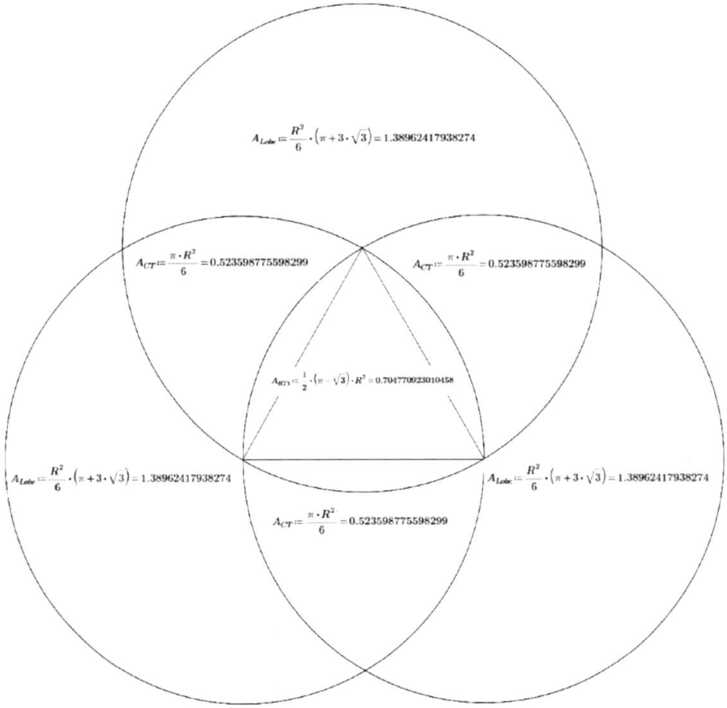

Figure 1.4
Component Area Equations

Corroboration of the Component Areas using the GIMP 3 Area Estimation Tool

It is always wise to check your mathematics by calculating numerical results using three independent methods.

I am getting idle in my old age and I shall content myself with only a second computation: I shall use the GIMP 3.0.2-1 graphics processing software to estimate the pixel population in elements of a partitioned trefoil; and move forward to compute the relative areas based upon R is unity.

Figure 1.5 is the screen image of a trefoil diagram loaded into GIMP 3.0.2-1. The four red annotations show points of interest when you use the Fuzzy Select Tool to estimate the area (in square pixels) of the click-selected central Reuleaux triangle. You can of course select any of the other six component shapes by clicking the Fuzzy Select Tool over them.

Table 1.1 shows the General Data Table used to cross-calibrate the area in square pixels with units based upon the ET side length R being 60 arbitrary units, whilst Table 1.2 presents the GIMP and Computed Areas for R = 60.

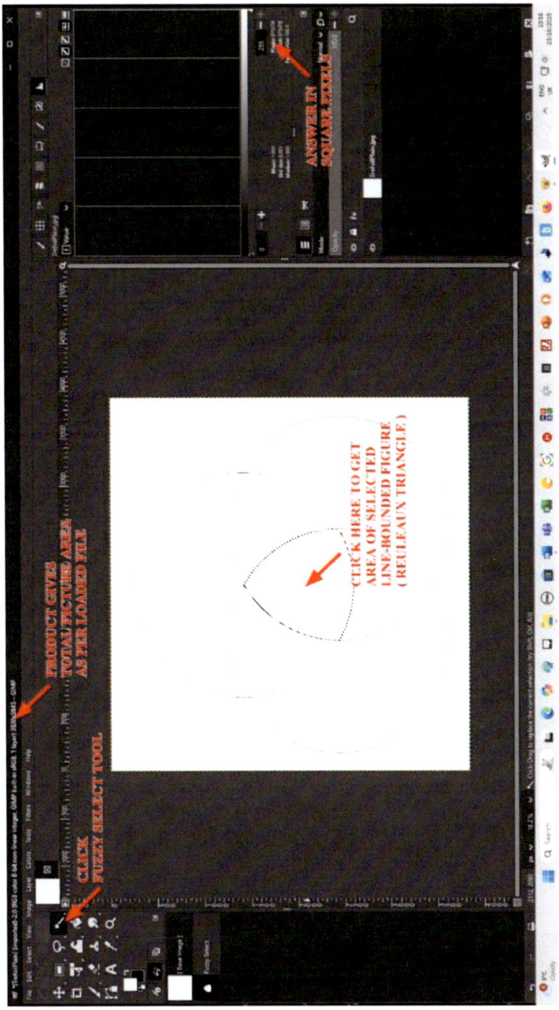

Figure 1.5
GIMP Area Estimation Screen

File:	TrefoilPlainHasMinorSegments.jpg
Image Datum Length (mm):	60
Datum Top GIMP x:	1952
Datum Top GIMP y:	1419
Datum Bottom GIMP x:	2546
Datum Bottom GIMP y:	2442
Datum Length (pixels):	1182.94759
Datum pixels/mm:	0.050720759
Datum pixels^2/mm^2:	0.002572595
Image Pixel per mm (x):	11.87113061
Image Pixel per mm (y):	11.81114456
Mean Pixel^2/mm^2:	140.2116398
Image Width (pixels):	3950
Image Height (pixels):	3845
Image Width (mm):	332.74
Image Height (mm):	325.54
Field Pixels (pixels^2):	15187750
Field Area (mm^2)	108320.1796
Field mm^2/pixel^2:	0.007132075
One Inch (mm):	25.4
Pixels per Inch:	300
Pixels per mm:	11.81102362
Pixels per mm^2:	139.500279

Table 1.1
General Calibration Data Table

Species Serial	Component Figure	GIMP pixels per Figure	Number of Component Figures	Species Subtotal ($pixels^2$)	Species Subtotal (mm^2)	Field Species Subtotal (mm^2)	MathCAD Value	Specific Defect (%)	
1	Lune	722378	3	2167134	5575.159	15456.1633			
2	Central Reuleaux Triangle RT	973523	1	973523	2504.481	6943.239532	2537.175323	1.28861827	
3	Lobe	1922534	3	5767602	14837.71	41134.97289			
4	Blob of 3 Circles	8964600	1	8964600	23062.29	63936.20398	23199.98324	0.59351021	
5	Synthetic Blob	8908259				0			
6	Outer Image			0	0	0			
7	Synthetic Major LENS	1695901	1	1695901	4362.867	12095.29396	4422.130915	1.34016289	
8	Segment 1	120574	1	120574	310.1881	859.9428707	326.1098653	4.88232525	
9	Segment 2	121360	1	121360	312.2102	865.5486821	326.1098653	4.26227041	
10	Segment 3	119984	1	119984	308.6703	855.7349462	326.1098653	5.34776082	
11	Total Minor Segments	361918	1	361918	931.0686	2581.226499	978.329596	4.83078549	
12	Equilateral Triangle	597068	1	597068	1536.014	4258.334052	1558.845727	1.46462988	
13	Composite Epitrochoid	958986	1	958986	2467.083	6839.560551	2537.175323	2.76261257	
								2.97474176	**Mean**
	Circular Triangle CT	722378	1	722378	1858.386	5152.054432	1884.955592	1.40954263	
	Lobe	1922534	1	1922534	4945.902	13711.65763	5002.647046	1.13429687	
	Blob	8964600	1	8964600	23062.29	63936.20398	23199.98324	0.59351021	
								1.04578324	**Mean**

Table 1.2
GIMP Measurements Data Table

Percentage Specific Defect is defined in our usual way as:-

$$\text{PSD} = 100\left(\frac{A_{eqn} - A_{GIMP}}{A_{eqn}}\right)$$

Equation 1.26

where A_{eqn} is an Area computed using Our Equation and A_{GIMP} is the Equivalent Area estimated by the GIMP software.

As you can see the discrepancies vary from about one percent to worse than five percent, mostly due to my poor draftsmanship. And GIMP varies slightly between different runs: For example the Table 1.2 value of Reuleaux Triangle area is 973523 px^2, whereas in my 23 October 2025 run to generate the screen diagram it was 973476 px^2.

CHAPTER TWO
THE DISCRETE AND THE CONTINUOUS

Mathematicians tend to think that objects are either Discrete, that is to say occurring as collections of individuals such as stars or starfish; or Continuous, existing as monotonously unbroken entities such as bodies of water or the atmosphere.

Modern science teaches that matter and energy are intimately related, and that whereas both appear to be continuous they are actually aggregates of tiny quanta of energy or matter. The quantisation of energy emitted from atoms reflects the fact that electrons can only "jump" between discrete orbits of the nucleus within an atom, whilst the very atoms themselves are discrete individuals which, however, seem to be continuous matter when they gather in their trillions and their billions of quadrillions.

In some sense a Trefoil is a discrete figure that limits us to seven internal compartments in which we can place entities which may be discrete or continuous. Such entities may be aggregates of one to three primitive ingredients. It is viable to think of an eighth compartment labeled zero which is everything outwith the Trefoil.

A Trefoil is a means of classification and a simple aid to thought.

But if we wish to describe continuous mixtures the Trefoil is not necessarily optimal. As a helpful diagram we may prefer the Ternary Plot of the kind exampled below:-

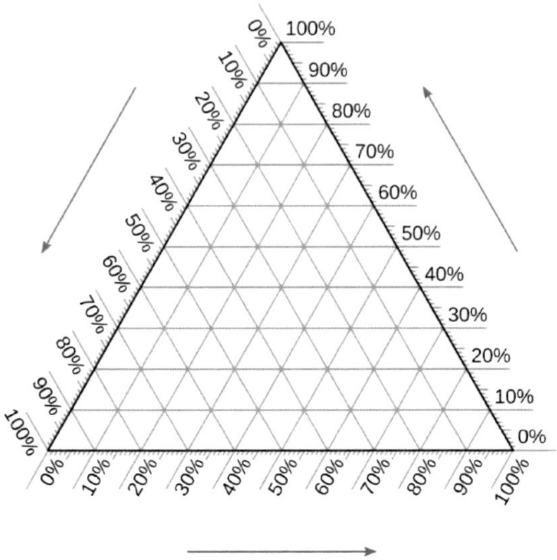

Figure 2.1
A Blank Ternary Plot

For example, we may have a building job that involves bonding bricks with cement-enhanced lime mortar. To be more specific, we may specify ASTM C 270 exterior load-bearing N-Type, mix of Cement:Lime:Sand in the ratios 1:1:6.

Now we are not sure whether this means the ratios are by dry volume or by dry weight: As a starting point we will assume the latter.

As an example of thinking involved we may look at the dry mixing of mortar for the bonding of a brick wall.

There are infinitely many ways of mixing sand, lime and cement with added water to make mortar, though few proportions are satisfactory.

One mix for cement-enhanced load-bearing external lime mortar is ASTM C 270.

Table 2.1 presents the dry densities of the chemicals referenced as well as their bulk densities as powder of which much or most is air.

Chemical	Substance Name in Building	Solid (Continuous) Density (kg/m^3)	Powder (Discrete) Density (kg/m^3)	Stated Proportions For ASTM C 270	Putative Mass Proportions	Putative Mass Percent	
Calcium Hydroxide	Hydrated Lime	2210	600	1	600	5.5198	
Silicon Dioxide	Sharp Sand	2650	1600	6	9600	88.3165	
Calcium Silicate	Portland Cement	2900	670	1	670	6.1638	
				8	10870	100	Sum
				2.666666667	3623.333333	33.33333	Mean

Table 2.1
The Possible Compositions of
ASTM C 270 Building Mortar

You can see that a good 88% of this mix is sand.

In real life Portland Cement is a mixture of tricalcium silicate, dicalcium silicate, gypsum, and sundry complex silicates involving iron and aluminum.

We could represent the dry composition of ASTM C 270 on a Ternary Plot as:-

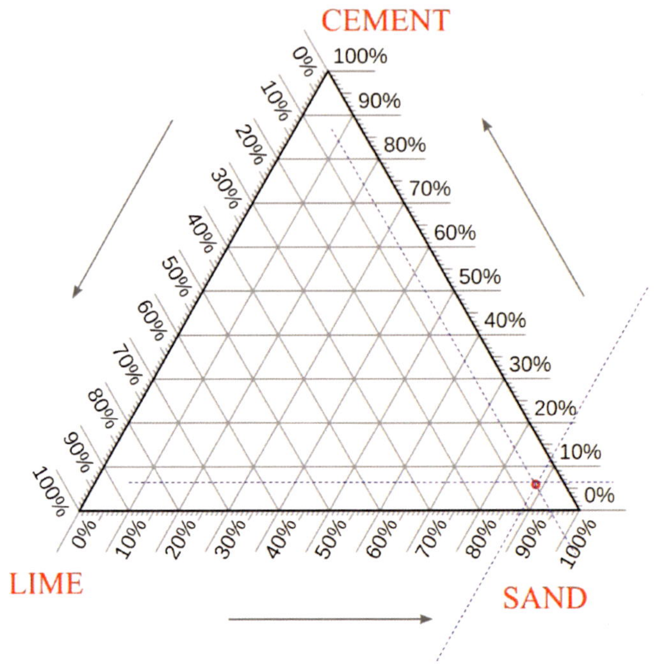

Figure 2.2
Ternary plot for ASTM C 270

The blue lines in Figure 2.2 denote the percentage levels of each elementary component and should intersect at a unique triple point in the center of the red circle. You can see that any. poor draftsmanship will lead to a zone of uncertainty around the point. Any continuous, analog, measurement is uncertain, in part due to errors of method and estimation.

In contrast, the partitioned Trefoil offers a discrete, finite representation of qualitative entities.

For example, Figure 2.3 allows us to depict the pure components in each of the three lobes and discriminate their mixtures in lenses and the central Reuleaux triangle, where the desired composite resides:-

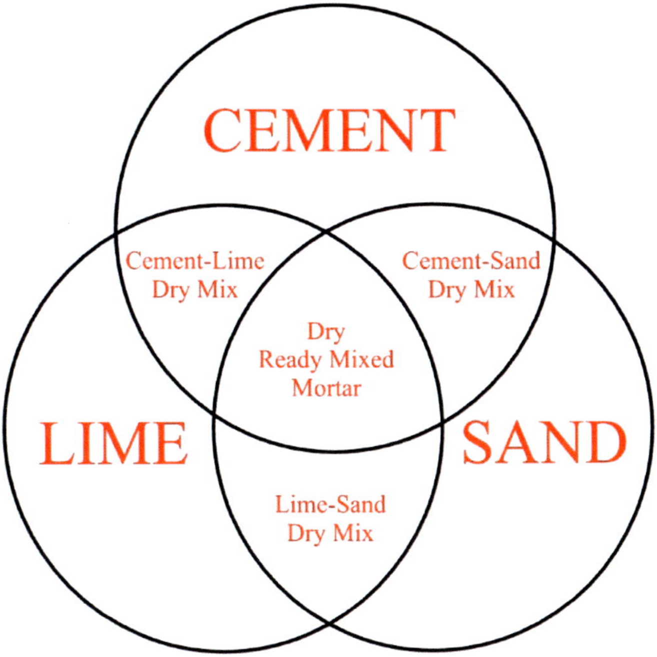

Figure 2.3
Trefoil Plot for Mortar Dry Mixes

It is readily apparent that a trefoil plot is clearer, more general, and more forgiving of my appalling draftsmanship!

In summary we may list the contrasting attributes of the Discrete and the Continuous in Table 2.2:-

ANALOG ALGEBRAIC	DIGITAL
Intrinsic	Extrinsic
Continuous	Discrete
Natural	Unnatural
Infinite	Finite
Quantitative	Qualitative

Table 2.2
The Discrete and the Continuous

CHAPTER THREE
DIGITAL EXPRESSION OF ASSOCIATION

In the Trefoil association is expressed by the overlap of internal zones as Lenses of Mutuality and their subdivisions. In this sense, a Trefoil is like a simplified Venn diagram. Various schemes could be devised for the quantification of mixtures in a Trefoil Plot, but the Ternary Plot is better for this and quantifications would sacrifice much of the clarity of the Trefoil diagram.

Because the Trefoil has only eight partitions (including its exterior) it is possible to label each compartment with a three-digit binary number as in Figure 3.1.

Note that the three circular triangles or Linguoids between the the primary Lobes hold the binary additions of the Lobe labels. This expresses the association of the primaries directly.

Meanwhile, the central Reuleaux Triangle combines all three primaries and is hence binary 111.

Something is missing.

The missing element is 000, which stands for everything outwith the trefoil, so that if the trefoil illustrates the compositional options of dry mortar, then 000 stands for everything which is *not* dry mortar.

So a more complete picture is that of Figure 3.2

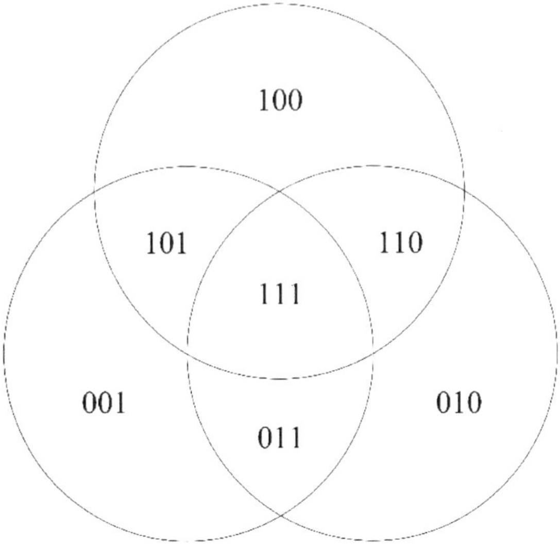

Figure 3.1
The Labeling of Trefoil Components with Binary Numbers

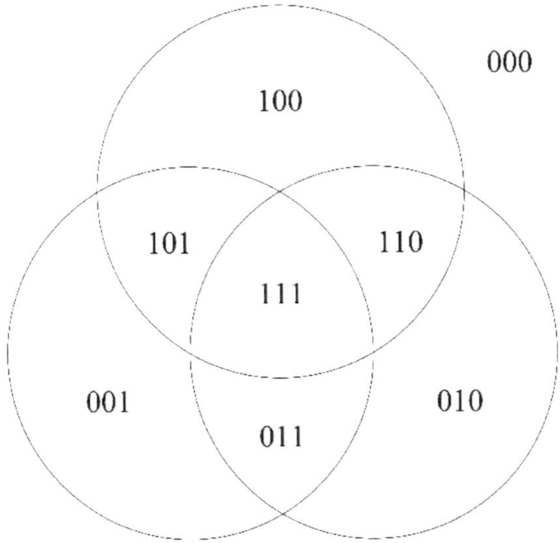

Figure 3.2
The Labeling of Trefoil Components with Additive Binary Numbers Including All Outwith

For those like me who are happier with denary digits, this is the denary equivalent:-

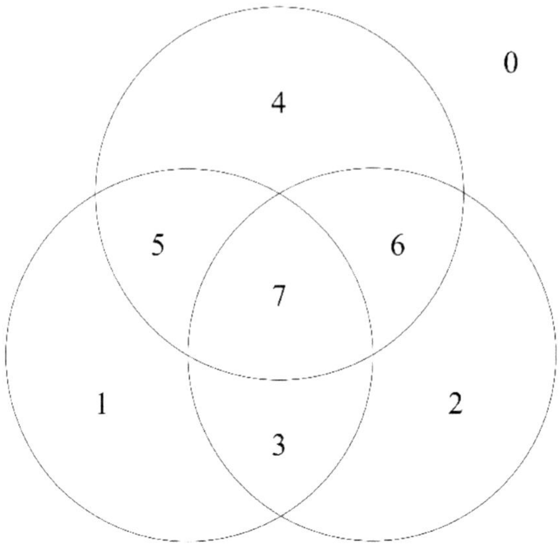

Figure 3.3
The Labeling of Trefoil Components with
Additive Denary Numbers
Including All Outwith

Additive and Subtractive Associations

Note that if you add opposing Lobes and Linguoids the result is seven in all three pairings, whilst each Linguoid bears the sum of the Lobes which overlap it.

The sum of the three Lobe labels is also seven.

By subtracting the values of adjacent Lobes we arrive at a different picture.

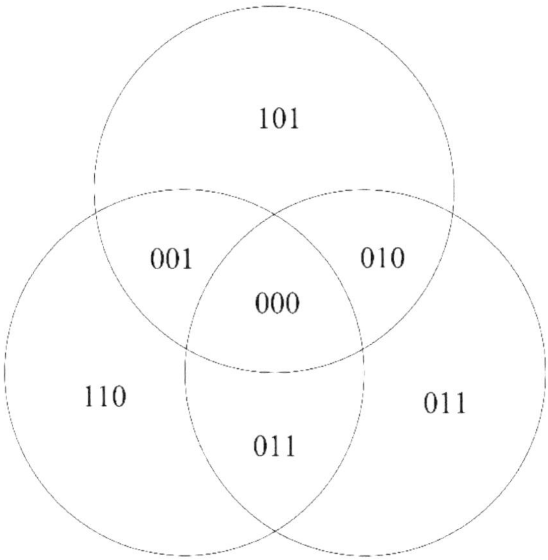

Figure 3.4
The Labeling of Trefoil Components with Subtractive Binary Numbers

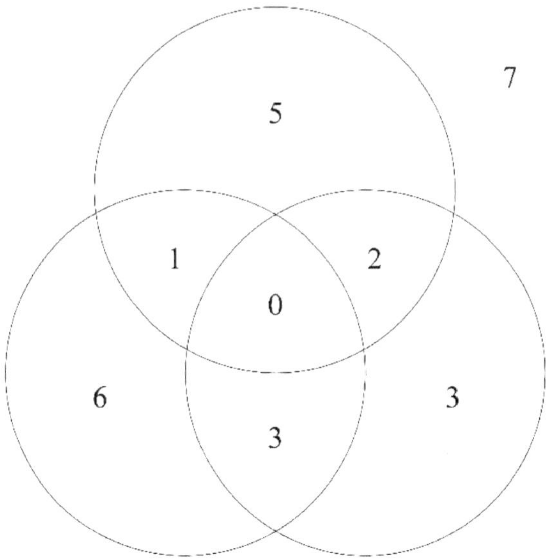

Figure 3.5
The Labeling of Trefoil Components with Subtractive Denary Numbers Including All Outwith

The *differences* of adjacent Lobes characterise the shared Linguoids.

CHAPTER FOUR
RELIGION

Since the days of Aristotle at least, philosophers, rhetoricians and strategists have canvassed a Rule of Three on the alleged basis that the human mind can grasp three principles and process them whereas two ideas are feeble and more than three confounding.

The Rule of Three has long fascinated politicians and journalists as evinced by such rhetoric as "Life, Liberty, and the pursuit of Happiness" from the United States Declaration of Independence or "of the People, by the People, for the People" from the Gettysburg Address. However noble the underlying sentiments such maxims seem often to bear a baleful and meretricious tone that belies intent.

Aristotle promulgated the Trivium, a system of thought and learning in three parts: Grammar, that governs the technicalities of entity definition; Logic, that governs valid operations upon information about entities; and Rhetoric that treats of teaching and persuasion. Within logic, the Ancients emphasised the utility of Syllogism, a tripartite structure that employs two valid independent predicates to define a third, supposedly ineluctable, consequence.

We have to be careful not to overstress such arcana, recognising their mnemonic usefulness, but also their tendency to the superficial and the trite.

It is in a liberal spirit then that we should review the trefoil, exploiting its illustration as an aid to thought but not allowing ourselves to be hidebound or obsessed by a mere image.

Body, Mind and Spirit

Mortal life is impossible without a Body which procures and sources the energy and structures upon which the Mind depends. The Mind is the essential creative principle which inhabits the Body of a creature, supporting the intellectual and elective powers that continuously optimise its probability of survival. The Spirit inhabits the Body, but when the Body fails the Spirit detaches and migrates elsewhere, because the Spirit cannot die.

Note that once again we are frustrated by what the English language is capable of expressing, not to mention the obvious mental limitations of this author. So for instance when we say “elsewhere” we intend no chorological implication, only meaning that the Spirit is independent of Body and Mind.

Greco-Roman thought and its current Western development posits that a heathy person, likely to survive long, balances his Mind, Body and Spirit in equipoise, or to put it negatively, dearth or disease of any of the three is likely to hasten death of the individual.

We must concede, however, that the three principles, if they exist at all, can exist independently or as couplets. One of the ways in which the interrelations of Mind, Body and Spirit may conceptually be organised is Penton’s System tabulated below:-

Denery Serial	Binary 2^2 Mind	2^1 Body	2^0 Spirit	Agent
0	0	0	0	Oblivion: Void
1	0	0	1	Passive Ghost
2	0	1	0	Corpse: Vegetable
3	0	1	1	Idiot: Lower Animal: Elemental
4	1	0	0	God
5	1	0	1	Angel: Demon: Satan
6	1	1	0	Computer: Psychopath
7	1	1	1	Christ: Higher Animal

Table 4.1
The Penton Ontological System Summarised

When you introduce ideas that are controversial or merely daft it is always an excellent idea to ascribe them to someone else, avoiding if practicable anyone you know to be alive. After all you do not want to receive rude letters or, worse, writs. So always cite someone fictional. Desiderius Penton is a character in one of my novels who bores an atheist friend with a very similar scheme.

A note for my younger readers: It is always very wicked to blame other people, even imaginary ones, for your own errors and crimes. Always own up to things you author or originate. You will get bouquets as well as brickbats, and even enemies will acknowledge that you are a real man or real woman. After all, we are not politicians or criminals who blame one another for our evil, and neither are we the projective and immature person who beats his teddy bear to assuage his own guilt.

So in terms of the Penton System God is pure Mind, a Creative intent whereas a passive ghost of the kind who is sometimes felt or seen in old premises is pure Spirit. Other entities are amalgams of two or more principles. For example, angels including Satan comprise Mind and Spirit but are incorporeal, whereas modern computers and psychopaths are soulless but very present in Body and intent. Both computers and psychopaths are endued with Evil by their human originators. Normal animals and the Risen Christ have Mind, Body and Spirit.

A handy mnemonic in terms of the trefoil might be:-

The Holy Trinity

The Holy Trinity is a concept at the core of Christianity, a concept of unity and diversity of three Agencies in one.

It comprises God the Father; Christ the Son; and the Holy Ghost.

All things were Created by God. Men and Women were created by God, but God saw that they erred and was displeased. So he begat an only Son and sent him to redeem humanity. Evil depends both upon intent and Agency, so non-human creatures, though frequently brutal, are incapable of Evil. The Holy Ghost is an Agency that, so to speak, communicates the Will of God and facilitates His Ministries.

Men stand upon the Earth and look upwards. Their transient and material condition necessitates that they are prone to Paradox, because their Knowledge is defective.

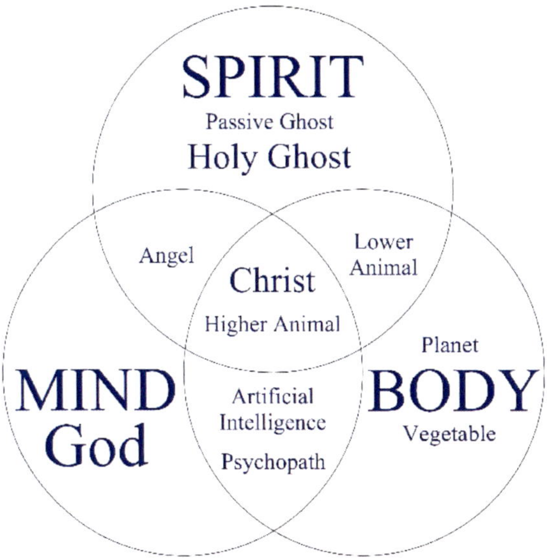

Figure 4.1
Penton's System in a Trefoil

Hence the often-cited syllogism, a great stumbling-block for many:-

God is all-powerful.
God is all-loving.
Evil exists.

It is sadly possible to infer that God is Evil, though love is not *in itself* evil. But it is more realistic to conclude that our Knowledge is incomplete and that the syllogism is logically defective. An evil God is belied by the fact that God sent a Man, *any* man, not an Oriental prince or a Norse hero, to save humanity from itself.

We may, as a matter of artistic taste or convention, assign primary colors to each of God, Christ and the Holy Ghost.

For example, a poet or a painter may represent God by the color red. Red is vital, pregnant and active: The hot lifeblood of Creation mirrored by the stars. Christ might be green, fecund and earthly like the fields of England or some other fertile and yieldful tract of land. Could the Holy Ghost be anything other than blue? Blue like the thin and sustaining sky, His cool and comforting natural home.

I am not much of an artist, but I offer the little triptych of Figure 4.2.

Iconography is the artistic science of representing philosophical concepts by means of coded images.

God is not of course an old man but like many others I represent him as a serious and concerned old man. He holds the World in his hands in my somewhat vulgar and literal illustration, wondering what to do with it, or for it. He rests within a pool of red, Number One and primogenitor of all, including the other two Agents.

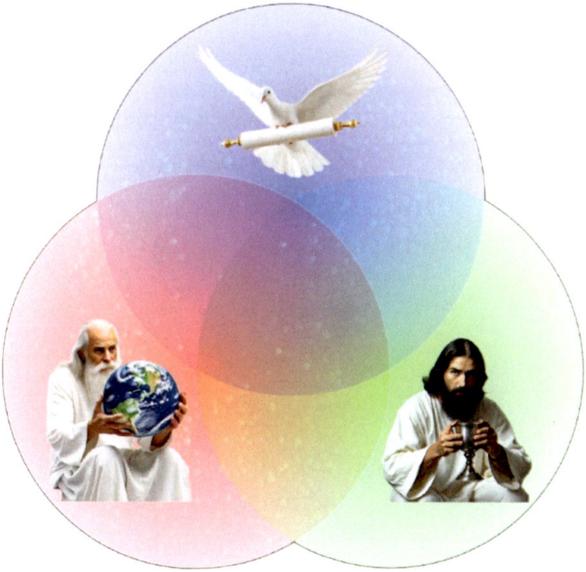

Figure 4.2
The Holy Trinity in a Trefoil

Number Two is Christ the Son, elect of God for the Purpose of Salvation. He too is very worried as he holds the Cup of Sacrifice and begs his Father to take it from Him (Matthew 26:39, Mark 14:36, Luke 22:42). What man wishes to be crucified? Jesus was a man: Christ was the Son of God the Father: An exacting Father who would welcome a triumphant Son home at last.

Christ rests upon a fruitful field of green, emblematic in my cod iconography of the yielding earth, corruptible but uncorrupted.

Number Three is the Holy Ghost. How can I represent something few have seen and many imagined? I think you would agree that it would be impossibly tasteless for me to signal the Holy Spirit with a picture of a Halloween bogy draped in a diaphanous shroud punctured by glowing red eyes. So I shall follow the others by representing Him by a white dove on the wing, ascendant and descending, with perfect control and inexhaustible strength. It is not helpful to imagine the Holy Ghost as a mere messenger or go-between: Rather he is equal and coeval, fulfilling his own essential Rôles. Remember that Godhead is a Unity.

The Ghost bears in his grasp the Book of Knowledge. It is an ancient book (*liber*) in the form of a scroll of serial text that must be unrolled to be read. Note that He brings it closed. Its content is not divulged to mortals.

The Linguoids of the overlapping circles are tinted with the mixed colors of the contributive elements, but the central Reuleaux triangle is a kind of tawny subfusc: **For now we see through a glass, darkly; but then face to face: now I know in part; but then shall I know even as also I am known.** (1 Corinthians 13:12)

A Six, another Six and then Another

Like many geometrical figures the Trefoil has some strange arithmetical properties that we may evaluate.

For example, assign the whole number 1 to God the Father as in the pattern of Figure 4.2; number 2 to Christ and number three to the Holy Ghost as below:-

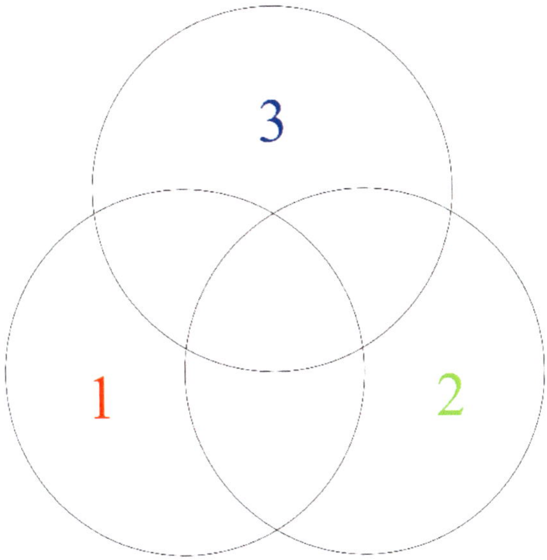

Figure 4.3
Assignment of Whole Numbers to Agents of the Trinity

This we may call this Stage A of our development and we have color-coded our elements as in Figure 4.2 for clarity and convenience.

In Stage B we add pairs of Agents numerical values and place the sums in the shared Linguoids as shown in Figure 4.4:-

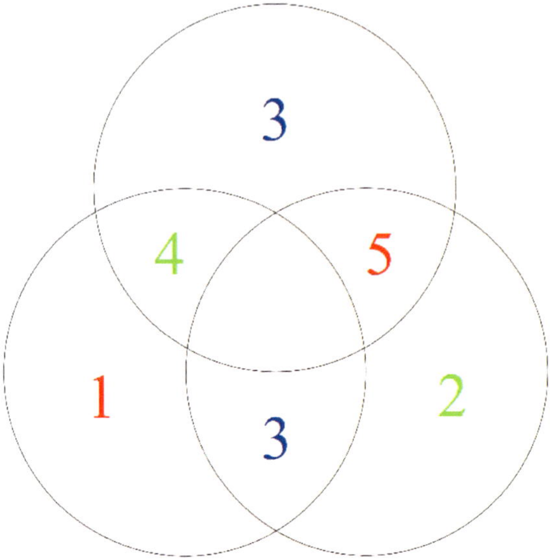

Figure 4.4
The Addition of Trinity Labels

In Stage C we add the Agents Lobe values to their opposite Linguoids and enter the totals in adjacent corners of the central Reuleaux Triangle. We color-code the sums for clarity but reduce the saturation of the Reuleaux secondary sums:-

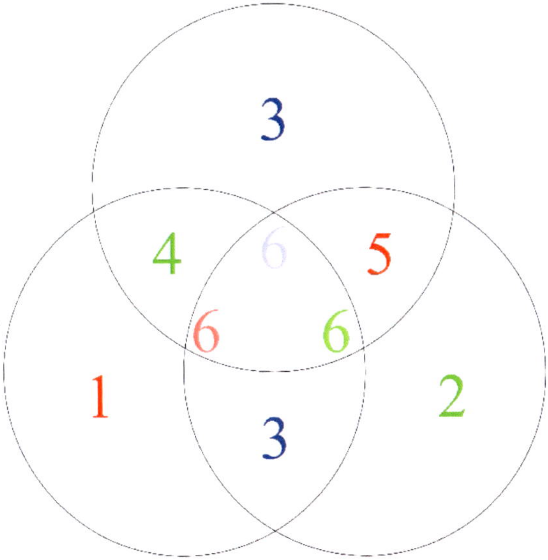

Figure 4.5
The Secondary Sums

We now have 666 in the center.

Note that this is *not* six-hundred and sixty-six, thirty-three score and six or any such unitary total: Such would be an anachronistic interpretation of 666 in Indo-Arabic positional notation, a concept alien to the Greco-Roman first-century (even that we designate retrospectively) the time which John of Patmos, author of the Book of Revelation, inhabited. The West did not learn of positional notation until the twelfth century AD.

The Greek way of representing this sequence of three digits would have been χξς (*not* positional) or ἑξακόσιοι ἑξήκοντα ἕξ (positional value of 666), and the Latin VIVIVI (not positional) or DCLXVI (positional value of 666).

VIVIVI is not standard Latin but may have been a colloquial form of the English expression "very much alive".

There are six possible permutations of the number sequence 1,2,3 and this digit triplet is unique in that the sums and products of each of the six arrangements are all six: This is true of no other triplet of numbers.

Table 4.2 illustrates this fact:-

First Digit:	1
Second Digit:	2
Third Digit:	3

Serial	**Permutation Position**			**Sum**	**Product**	**Quotient**	**Difference**
	1	**2**	**3**				
1	1	2	3	6	6	0.166667	-4
2	1	3	2	6	6	0.166667	-4
3	2	1	3	6	6	0.666667	-2
4	2	3	1	6	6	0.666667	-2
5	3	2	1	6	6	1.5	0
6	3	1	2	6	6	1.5	0

Table 4.2
The Sums and Products of the Digits 1, 2 and 3
In All Possible Orders

To those who consider that socialism, national or international, or some other ideology or some anthropogenic system such as mathematics or science or something, or who advocate National Identity Cards, physical or figurative, are the solution to any problem I commend The Book of Revelation by John of Patmos.

You can find it at the end of any Christian Bible.

Read Revelation 13:13-18:-

And he doeth great wonders, so that he maketh fire come down from heaven on the earth in the sight of men,

And deceiveth them that dwell on the earth by the means of those miracles which he had power to do in the sight of the beast; saying to them that dwell on the earth, that they should make an image to the beast, which had the wound by a sword, and did live.

And he had power to give life unto the image of the beast, that the image of the beast should both speak, and cause that as many as would not worship the image of the beast should be killed.

And he causeth all, both small and great, rich and poor, free and bond, to receive a mark in their right hand, or in their foreheads:

And that no man might buy or sell, save he that had the mark, or the name of the beast, or the number of his name.

Here is wisdom. Let him that hath understanding count the number of the beast: for it is the number of a man; and his number is Six hundred threescore and six.

Now all Bibles say this number as a *unitary total* being six hundred and sixty-six items, though they might print six hundred, three score and six or some other equivalent formula. But think about what John has actually written. The number is 666 which is a *sequence* of three specific digits. It is our habit that construes it a sum, about two-thirds of one thousand.

John was trying to tell us, in his own idiosyncratic manner, that just as six, another six and a companion six inhere in a Divine Trefoil: Just so, Evil inheres in *everything*. If we choose to survive, then it is our task to unmask evil.

CHAPTER FIVE
COLOR

White light, as perceived by the human cognition, is an amalgam of several more or less distinct colors that form a visible spectrum. Children of England are taught that "Richard Of York Gained Battles in Vain" as a mnemonic for "Red, Orange, Yellow, Green, Blue, Violet" to give us the order of colors in the rainbow's spectrum: An order of increasing energy.

Perceived light may either be Transmitted through matter or free space, or else Reflected by matter. If the matter is translucent, then it will reflect some light and transmit or absorb the rest. Energy absorbed is often realised as heat.

The sensible Composition of Light differs according to whether it is Transmitted or Reflected.

Darkness is the dearth of light and blackness its absence.

Transmitted and Reflected light are key tools in the analysis of minerals and of rock fabric.

When I remember this I cannot help remembering also the divine verse of George Herbert (1593-1633) from his poem "The Elixir" of 1633:-

A man that looks on glass,
On it may stay his eye;
Or if he pleaseth, through it pass,
And then the heav'n espy.

So true.

Figure 5.1 represents rays of light Transmitted and Reflected from a sheet of glass. The red dotted line is Reflected light from the first interface of air and glass: The blue dashed line Transmitted light which crosses the glass to regain air.

Note that the Transmitted light is *refracted* (changes direction) in the optically-denser medium and regains its original trajectory on re-entering the air.

Primary colors are three essential colors that can be mixed to produce an infinite variety of other hues. The Primary Colors for Transmitted light are Red, Green and Blue. Transmitted Light colors are Additive Colors. When red, green and blue beams of light mix they produce white light, as in theatrical filtered spotlights.

The Primary Colors for Reflected light, such as that of mixed artist's pigments, are Red, Blue and Yellow. But in the case of these colors red, green and blue and indeed any other color, can be made by mixing cyan, magenta and yellow, just as is done by printing machines upon paper. These are Subtractive Colors. When cyan, magenta and yellow are mixed as a pigment they theoretically produce black, though in practice the result is a murky brown or tawny paste. It is better to use a manganese or carbon black if a pure black is what you want.

As stated, red, green, blue and yellow can be combined in any proportions to produce the exact color desired, and these basic Primary colors are defined in Table 5.1, but note that they may be a little "off" in the rendering of your page.

In the composition of computer or printer hues it is usual to define the proportions of primary colors on scales of 0 to 255, because in the binary system 2^8 is 256, whilst 8 is 2^3.

The Additive Red-Green- Blue color system is known as the RGB System: And the Subtractive Cyan, Magenta, Yellow color system as CMY, or CMYK when supplemented by a pure black pigment in order to suppress muddiness.

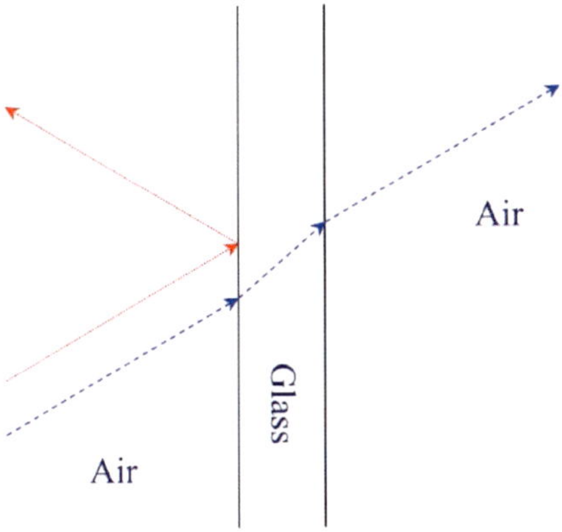

Figure 5.1
The Reflection, Refraction and Transmission of A Ray of Light

System		R	G	B	Binary Code	Realisation
ADDITIVE	Red	255			100	
	Green		255		010	
	Blue			255	001	

		R	G	B	K		
SUBTRACTIVE	Cyan		255	255		011	
	Magenta	255	0	255		101	
	Yellow	255	255	0		110	
	Black	0	0	0		000	

Table 5.1
Primary Colors Defined

Note that binary numbers can be assigned to label every primary color and these binary labels can be used to relate each color to a zone of the Trefoil.

Figure 5.2 shows Trefoil zones in Additive Primary and Mixed colors. Note that the Primaries characterise the outer Lobes and the Mixed overlapping Lenses.

Figure 5.3 shows Trefoil Zones in Subtractive Primary and Mixed colors. Note that the Primary Cyan, Magenta and Yellow are in the outer Lobes, and the derived Red, Blue and Green in the Lenses:-

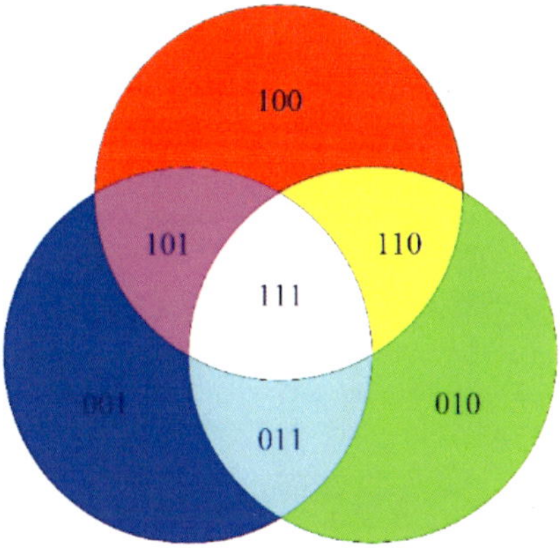

Figure 5.2
The Addition of
Transmitted Colors

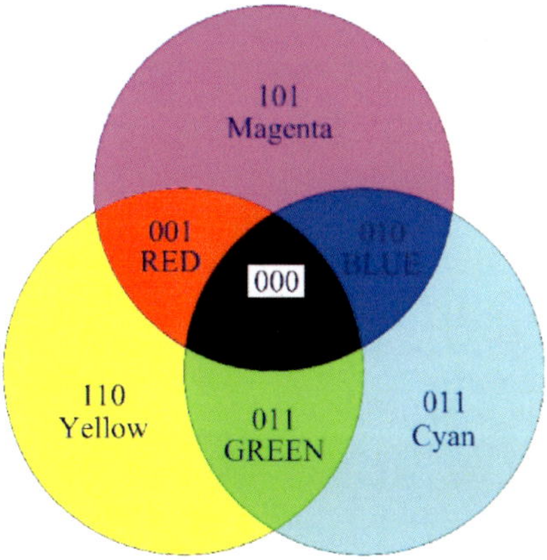

Figure 5.3
The Subtraction (Absorption) of
Reflected Colors

Glass

I well recollect a school assembly of 1957 at Littlehampton. I would have been five. First thing every morning, it was the teacher's practice to troop her little charges to the assembly hall of the Maud Allen Girls' Secondary Modern school next door. Before proceedings every morning she would play the same Mozart theme to us on a Dansette. The morning sun, diffused by sea-mist, streamed through the wide plate glass windows.

The teacher stood at a lectern and we children sat cross-legged on the wooden floor that stank.

Shortly before my Nan had told me, apropos what I still know not, that glass was made of "sand, soda and lime".

For some reason I cannot remember, perhaps inspired by the expanse of sparkling glass, my teacher, I seem to recollect the egregious Miss Calcutt (married women were not allowed to teach in England in those days) asked the sixty or so of us what glass was made of.

I raised my hand and shouted "Sand, Soda and Lime, Miss."

The other children dissolved in paroxysms of laughter.

Our teacher shouted back: "Yes, that is right, Jimmy."

The others gasped in incredulity.

To the other children sand was sand, familiar from the beach, and soda and lime were fizzy water, delivered, if affordable, by the Corona man.

For myself I had little concept of what soda and lime were but Nan had in mind Sodium Carbonate Na_2CO_3, common washing soda (manufactured detergents were rare and expensive at the time); and ground limestone, Calcium Carbonate $CaCO_3$. Nan had undoubtedly learned this from the

Black Country glassmakers whom she and her sister had befriended as girls in the reign of Victoria.

I think Miss Calcutt was as surprised as the children: Not at the composition of glass, but because she regarded me as the class idiot, unable to wrangle the Imperial System of Weights and Measures. In fact I was demoted to the dunce's class for this latter failing.

Neither the Littlehampton County Infants' School nor the Maud Allen survived the twentieth century and their buildings were demolished. Their sites are now (2025) occupied by housing. The Imperial System of Weights and Measures was effectively abolished between 1963 and 1985 when the pound and the gallon were defined in Metric Units. I made a point of never using the Imperial System.

It was a different world, and children are rightly treated much better in contemporary Britain, and often taught silly things, but for their personal benefit rather than that of the Empire.

Colored glass is anything other than colorless and transparent only by virtue of additives. Glass proper is, as physicists say, not a solid, but a "supercooled liquid" of very high but finite viscosity: This is shown by the gradual creep and basal thickenings of plate glass windows. As glass ages it may however crystallise to give a beautiful internal iridescence as often seen in Ancient Roman glass unearthed.

Dr Chloë Duckworth has described the proficiency of the Ancient Egyptians at colored glass making in the Late Bronze Age (1500-1200BC) where glass was exploited as a cheap substitute for precious gems like turquoise or lapis.

Several techniques may be used to color glass:-

A **Body Color (“Pot Metal”)**
The coloring agent, usually a metallic salt or oxide, is mixed into the melt

B **Staining**
A wash of a cold salt, often silver nitrate, is painted on cold plate glass which is then fired.
Silver nitrate will produce yellow to orange-red staining.

C **Flashing**
A thin plate of colored glass is welded to the surface of a glass base substrate, which may be clear.
This is favored where a thicker coat of pot metal would be too dark or even opaque.

Red Colors

Medieval red glass was often developed from mixing finely-comminuted gold (Au) into ordinary pot glass, but it was difficult to control the depth of color, which often appeared black at a pane thickness of only three millimeters. So a thin lamina of red stained glass was formed and welded (“flashed”) to a clear substrate. Gold content could be as low as one part in ten thousand, and the gold could be incorporated into tin-laced (Sn) lead (Pb) glass to produce rich red or pink coloration.

Red glass can also be made with comminuted copper (Cu) and this was used extensively in the twentieth century.

Modern red glass is stained with selenium (Se), sometimes enhanced with cadmium (Cd) sulfide.

Green Colors

Green glass can be produced using iron (Fe) (II) salts, and the color is made richer by the addition of chromium (Cr). This richer variety if suitable for flashing (the enameling of plate glass). Addition of tin (Sn) oxide and arsenic (As) results in emerald green.

Blue Colors

Medieval blue glass was achieved by adding 0.025 to 0.1% of cobalt (Co) blue (cobalt oxide or cobalt carbonate) to the melt.

Modern blue glass is developed by adding nickel (Ni), or in the case of borosilicate glass, sulfur (S).

Two to Three percent of copper oxide gives a turquoise color.

Yellow Colors

Pre-industrial yellow glass was often produced by staining with silver (Ag) nitrate as above noted, but modern glass is produced by the addition of sulfur and calcium (Ca). Cadmium (Cd) and sulfur yield a deep yellow, and uranium (U) can be used to impart yellow or green.

It cannot be guaranteed that the simple mixing of colorants will produce theoretically-mixed shades because the constituent chemicals of glass and its dyes inter-react.

Also, lead, arsenic, cadmium and uranium are all toxic and appropriate precautions must be taken in their use. Purple glass may be made by using manganese (Mn) or nickel (Ni) salts in the pot mix, whilst white may be made using antimony (Sb) and arsenic (As) compounds.

In a very rough and ready way we can illustrate the staining of glass using Figure 5.4 below:-

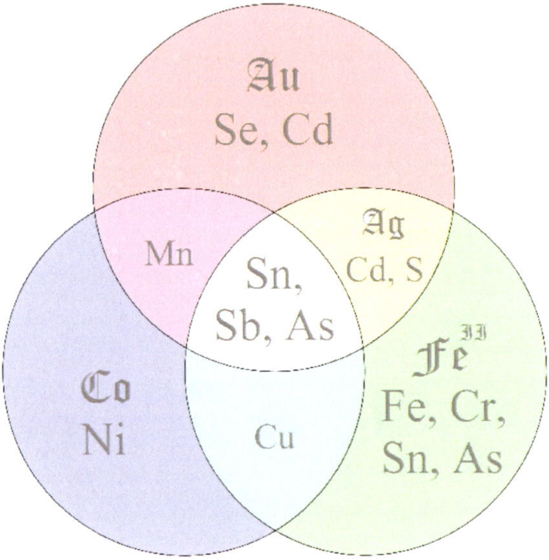

Figure 5.4
Chemical Elements to
Stain Glass

The Chemical Symbols in **Old English Script** denote staining additives employed in pre-industrial times, whilst Times New Roman script denotes modern materials.

The tints have no implication regarding color intensities. The colors are muted in order to aid visibility of the zonal chemical symbols.

CHAPTER SIX
ART AND ENGINEERING

King Tut's Bedsteads

Among the manifold wonders amidst the disjecta that Carter dimly beheld in the Autumn of 1922AD where two elegant gilded funerary bedsteads of Hathor, the gay goddess of love, sex and nurturing motherhood, standing as the mortuary comforter Mehet-Weret, "The Lady of the West".

The conceit, or belief, that the West is the destination of the dead is of extreme antiquity and probably universal. My maternal grandmother, a woman of Victorian Worcestershire, she of the sand, soda, and lime, habitually spoke of damaged or destroyed things as "going West".

In the Tomb of the boy pharaoh Tutankhamun (1341-1323BC), Hathor is represented in her form as a golden cow with black trilobate spots. Some have said that these represent the stars of the firmament referencing Hathor's alternate tutelage of the night sky, but to the Ancients the stars were little more than formless points of light, some of which moved and many of which stood still. Perhaps the spots of Hathor are more like the maculations of many mammalian hides, somewhat like the vaguely pentamerous spots of a leopard, or the mottlings of some modern cattle.

Figure 6.1 is a photograph of one of the said bedsteads taken on the opening of Tutankhamun's tomb in 1922:-

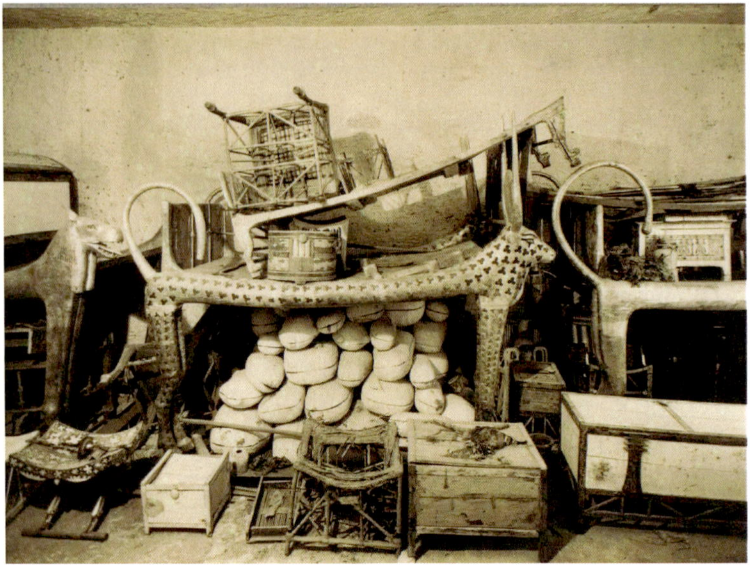

Figure 6.1
The Mortuary Effigy of Hathor as Guide of the Dead

Figure 6.2 offers a more distinct image of Hathor's head.

Clearly, the ancient craftsmen had a die or stamp with which they impressed the pattern, but the principles upon which they wrought the die or spaced its impressions are lost to history.

I have attempted a re-creation in Figure 6.3.

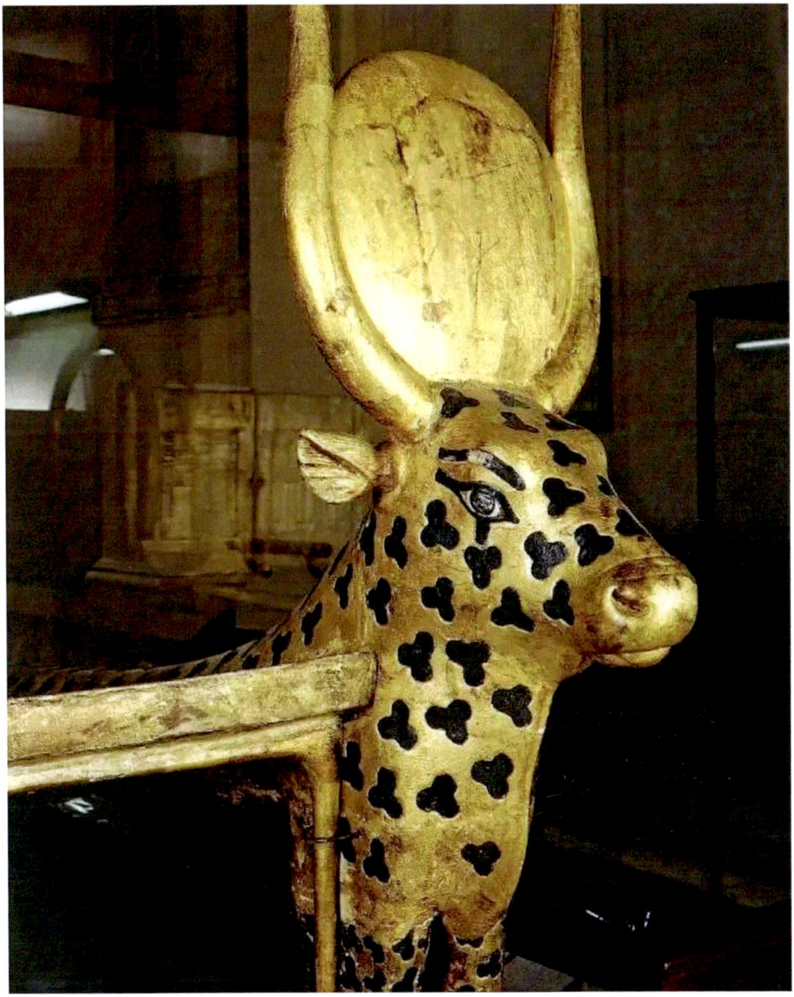

Figure 6.2
The Head of the Bovine Hathor
As Mehet-Weret
From the Tomb of Tutankhamun

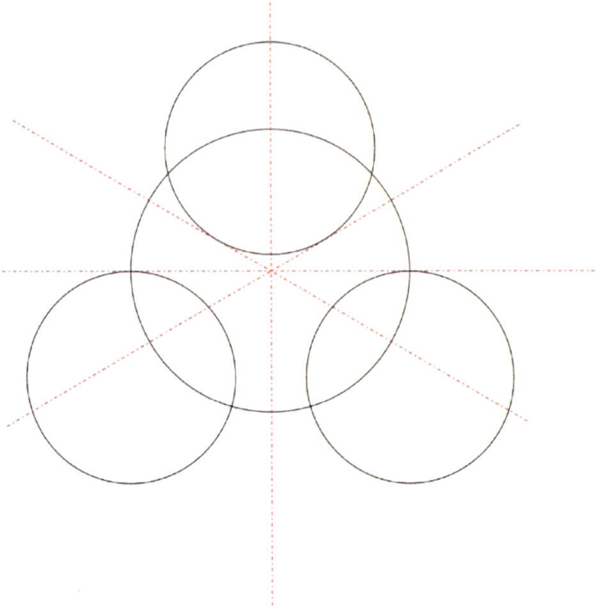

Figure 6.3
Author's Approximation of the Trilobates of Hathor

In this very simple design the radius of the central large circle is 200 units and that of the three outer circles 150 units, so that the diameters are 400:300 reducing to 4:3.

The subsidiary circles are aligned with the circumference of the principal and the circumference of the upper circle set on the radii at 60° to the center of the principle, central, circle.

The infilled result is illustrated in Figure 6.4:-

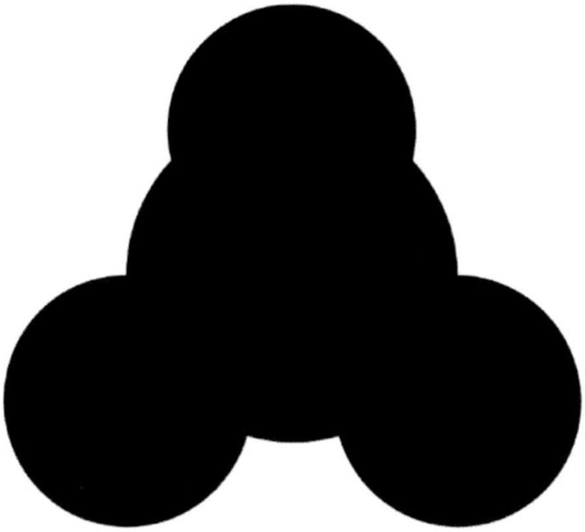

Figure 6.4
An Approximate Re-Construction of the Hathor Trilobate

This is not strictly a trefoil in our more sophisticated terms but it exemplifies the many constructions that were available to carpenters and masons of the Late Bronze Age working with *straight-edge and compasses only*.

The R = 0.25 Unconnected Trefoil

Suppose that in the terms of the Chapter One trefoil the circles' radii are not unity but 0.25 or any radius less than 0.5: Then the trefoil is undeveloped and the zonal overlaps non-existent, as illustrated in Figure 6.5:-

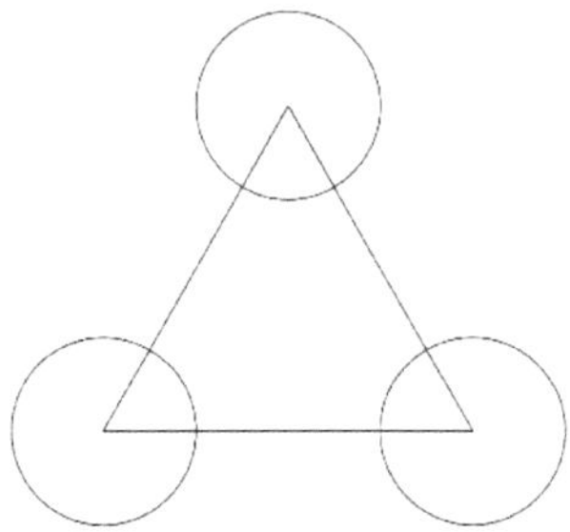

Figure 6.5
The Undeveloped R = 0.25 Trefoil

Some artists and craftsmen of the past have nevertheless given us such circles at the vertices of the equilateral triangle in the manner of Figure 6.6.

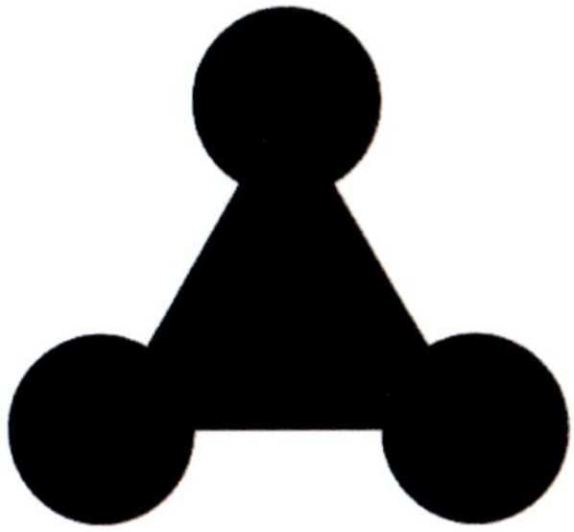

Figure 6.6
An Artistic Device:
Circles upon an Equilateral Triangle

The Osculating R = 0.5 Trefoil

In the case of the radii being half of the side of the equilateral triangle (R = 0.5), the three circles osculate (their perimeters just touch).

There is one internal zone and the generated trefoil is as illustrated in Figure 6.7:-

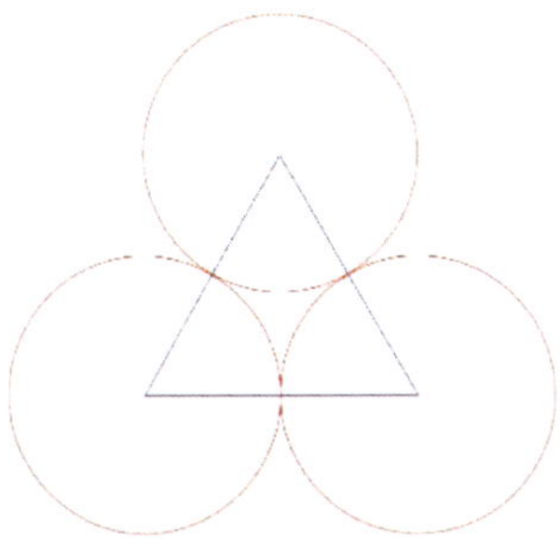

Figure 6.7
The Basic Osculative Trefoil

The $R = \phi$ Trefoil

If the common radii R have the value of the Minor Ratio of Phidias, ϕ, which has the approximate value 0.618034, then the figure is as Figure 6.8:-

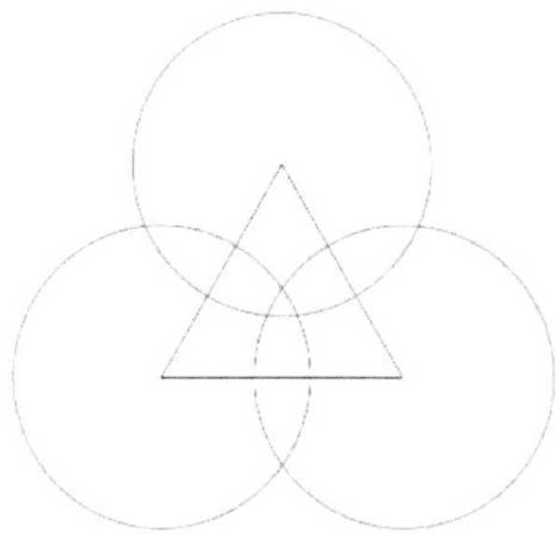

Figure 6.8
The Basic Osculative Trefoil with $R = \phi \approx 0.618034$

Types of Arches

The Trefoil, judiciously employed, can help us to design structures as well as organise our thoughts more generally.

All things which stand on a firm surface within a gravitational field bear load, whether from a pay-burden or their own self-weight or both. Among such structures are bridges, mills, bastions, revetments, cathedrals, bunkers and towers, not to forget cars, trains and ships, sheds or anything else built.

The walls and piers that support structures are seldom uniform because they are pierced with holes and windows to permit ingress and egress, or admit light and air. Such piercings (fenestration) introduce weakness to the wall because they interrupt the support of competent materials and introduce points of concentrated stress at the breaks in continuity, such as the corners of windows or doors.

Such breaks are subject to stress which is direct or re-directed weight, technically a pressure, expressed in terms of force per unit area.

There is vertical downward stress due to the weight of the arch itself and its supported burdens; and lateral thrusts outwards from the aperture, due to re-directed weights. In addition these vertical and lateral thrusts give rise to bending moments and torsions within the arch itself which can lead to failure and collapse.

The Trefoil has a rôle to play in the design of load-bearing arches, and not just as decoration.

There are four main kinds of classical arch:-

A **Architrave or Lintel**

This is a basic straight beam of ashlar or reinforced concrete like the straight megalith spanning the columns of an Ancient Greek temple or the sarsens of

Stonehenge. It is unsuited to long unsupported spans, due to the intense bending moments at the lintel center which tend to turn and shatter it like a bent stick.

Limestone and reinforced concrete are most suitable for lintels because both have a limited elastic flexibility.

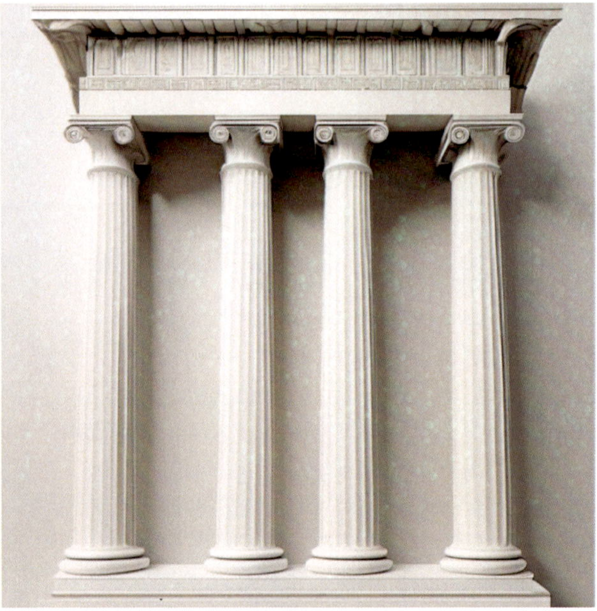

Figure 6.9
A Lintel Supported by Columns

Lintels can support limited courses of superincumbent ashlar and light rooves, such as the tiled wooden rooves of ancient buildings.

All burden stress is directed downward through the supporting piers (columns).

B **The Semicircular or Roman Arch**

This comprises a semicircle of crafted wedge-shaped voussoirs of stone either mortared or dry, and arrayed above temporary centering scaffolds to form a self-supporting semicircle, with or without a top central keystone.

This structure is capable of great stability, but is subject to high lateral, outward stresses which can incur tearing strain at intermediate points in the arch soffit.

The Roman arch is suited to the support of massive static loads in such structures as stone bridges, brick viaducts, crypts, sea walls and military fortifications, and the lower stories of buildings housing heavy plant such as mills.

Subterranean structures such as tunnels and sewers are nearly always capped by semicircular arches, these being natural to cylindrical bores.

The Ancient Romans and the Victorians both incorporated blind semicircular arches into curtain walls as reinforcement.

Semicircular arches tend to obstruct the ingress and egress of air and light.

The segmental arch is a special case of the semicircular arch in which the span of the arch is a chord of a circle rather than a diameter, giving a long, shallow profile. This configuration concentrates massive parting stresses at the abutments, and when the Ancient Chinese favored this profile for bridges over their canals, they found that extensive stone reinforcements were necessary beneath the bridge approaches. However, like many Roman bridges and aqueducts, these structures have survived until the present day.

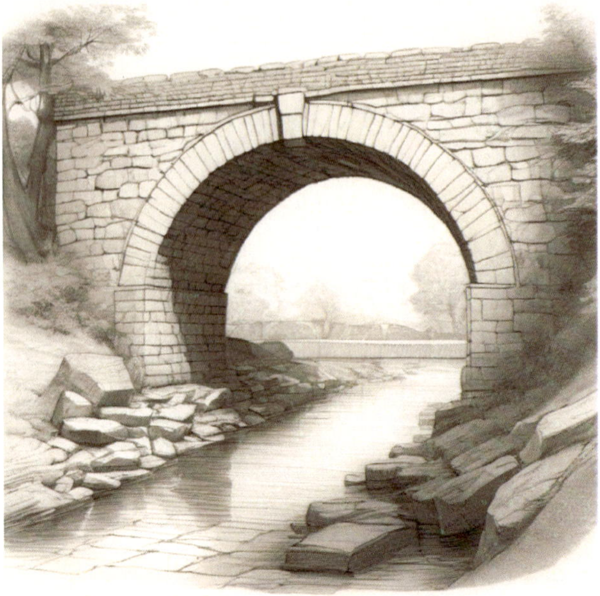

Figure 6.10
A Semicircular Arch

C The Cuspate or Gothic Arch

The Gothic arch has curved sides, typically arcs of circles, that come to a point at the apex, which is therefore a point of discontinuity and potential weakness. A favorable property, however, is that stress within the arch is more uniformly-distributed along the sides than is the case with a Roman arch.

Cuspate fenestration is vulnerable to lateral stresses and though cuspate arches permit walls to be built thinner and cheaper, lateral thrusts must usually be compensated by buttressing, including “flying buttresses”, which are thin arcuate bridges of stone which offer compensating inward stress to support a tall thin wall from outwith.

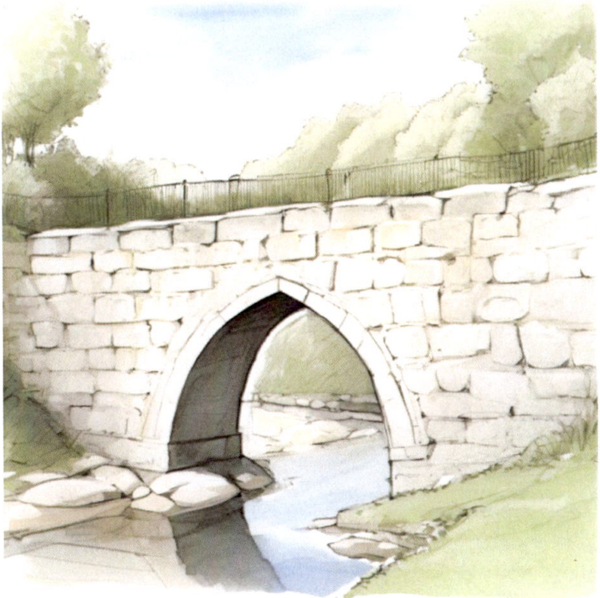

Figure 6.11
A Cuspate Arch beneath a Bridge

The Cuspate Arch in medieval cathedrals is not called upon to shoulder massive superincumbent loads: Only low parapets and wooden rooves waterproofed with thin plates of lead or copper.

D The Parabolic or Ctesiphon Arch

A parabolic arch has uniform stress distribution along the whole arc of the structure.

A symmetrical parabolic arch profile conforms to the second-order polynomial equation:-

$$y = \alpha x^2 + h$$

Equation 6.1

where y is the Local Soffit Height above the Parabola Base; α is a Form Constant; x is the Lateral Base Displacement between x-L/2 and x+L/2, where L is the Basal Span; and h is the (Symmetrical) Arch Altitude in the Center, the “Rise”.

For a Cuspate Arch, lateral stresses are minimised but they require massive foundations where the two arms enter the earth and thin walls require exterior buttressing as previously remarked for the Cuspate Arch, because a Cuspate Arch *approximates* a parabolic arch.

It is mathematically impossible for a parabola to *exactly* match a cuspate arch built from circular arcs, because all arcs of circles are generated from formulae involving the Ludolphine Constant, π, a transcendental number which is not the root of any algebraic polynomial, whereas a parabolic (i.e. a second-order algebraic) equation has, by definition, two roots which are the dividends of integers.

For God all things are possible, but for men few things are possible. Mathematics is a man-made artificial construct, a language, and like all the works of man incorporates his errors and limitations.

But for ourselves, as practical craftsmen, we can work with limits and possibilities to produce the best we can.

The Classical architype of the parabolic arch is the great arch, the Taq Kasra, at Ctesiphon in modern day Iraq. It is shown below in a photograph of 1864:-

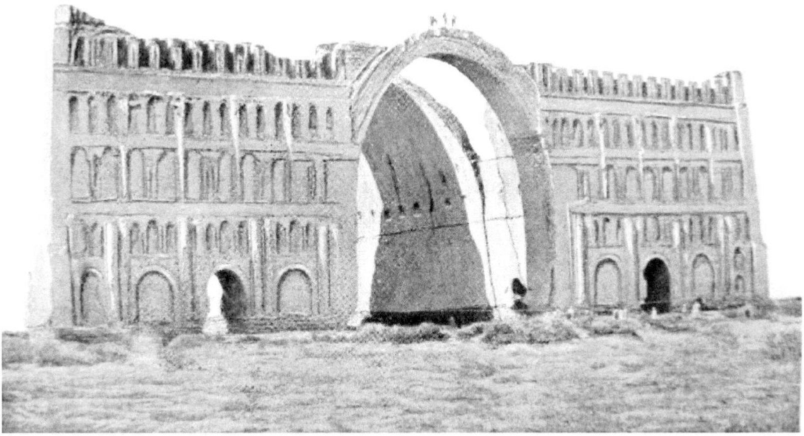

Figure 6.12
The Parabolic Arch of Taq Kasra at Ctesiphon

The Taq Kasra arch is thought to have been built in the Third Century AD by the Sassanian regime. The interior arch still stands though the facia arch bearing the three human figures has been lost.

In modern times the parabolic arch has sometimes been used to support road and railway viaducts, permitting relatively gracile steel or concrete structures as long as sufficiently robust foundations are available in rock or mass concrete. For example:-

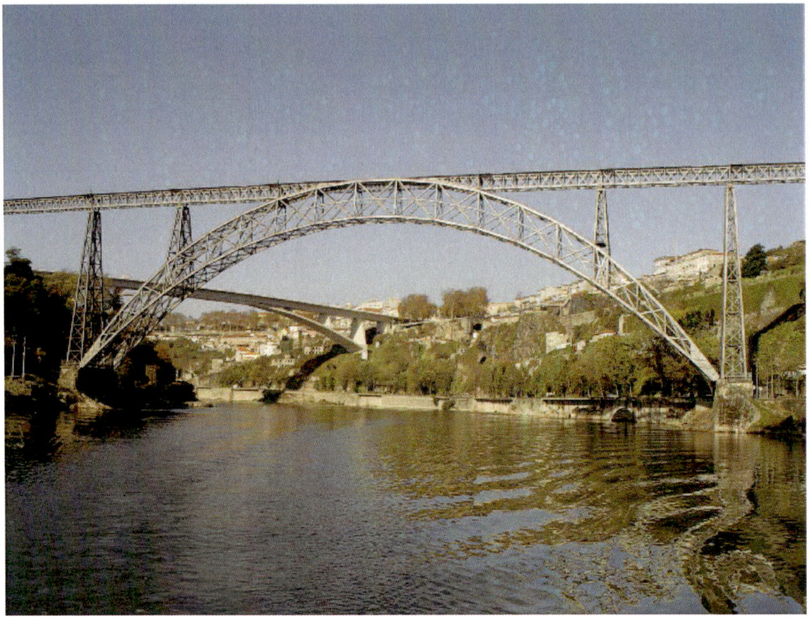

Figure 6.13
The Dona Maria Pia Bridge spanning the Duro at Oporto

The Dona Maria Pia Bridge in Portugal was built by Gustave Eiffel in 1877 and stands. Note the lightness of the rivetted steel latticework and the massiveness of the concrete foundations.

Designing a Lancet Window

A lancet window is a tall, thin window topped by a cuspate arch. Superincumbent loads are transmitted perpendicularly to ground through the jambs and mullions of the window structure.

The Trefoil Linguoid can be used to create the skeleton of the cuspate arch at the top of such a window, a strong compartment into which tracery can be fitted to accommodate stained glass.

First, a trefoil of the basic type is drawn with the common radii of unity. Any of the three available Linguoids is isolated, together with its basal unit-long base line from the central equilateral triangle, as illustrated in Figures 6.14 and 6.15:-

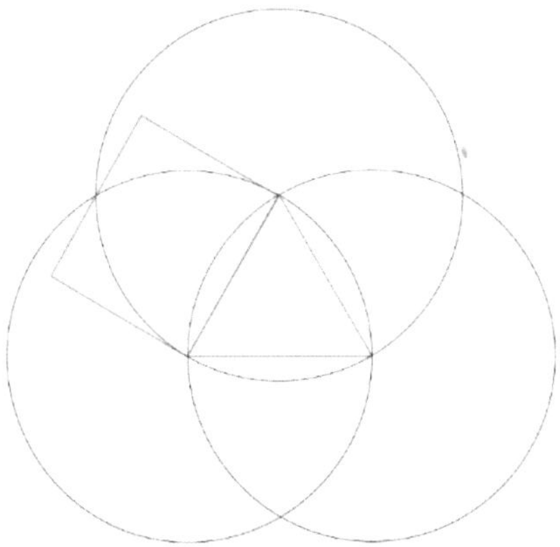

Figure 6.14
Trefoil with Linguoid Extraction Zone
As a Red Rectangle

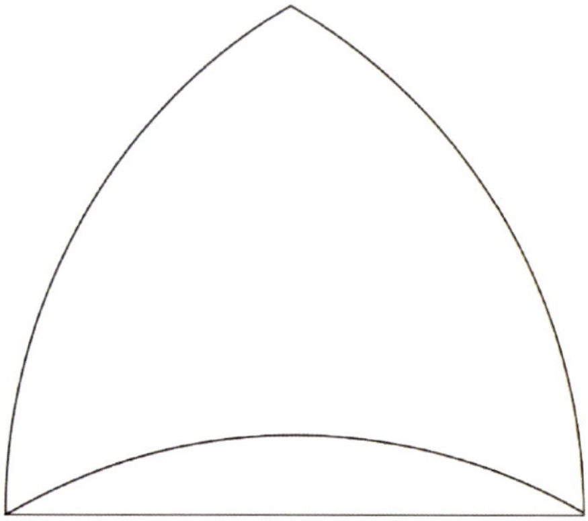

Figure 6.15
Linguoid Extraction Zone

Next we measure or calculate the rise and span (i.e. height and width) of the Linguoid as given in Table 6.1:-

Lineament	Symbol	Diagram Length (mm)	Length Base is Unity
Length of Basal Segment	L	100.1	1
Half Length of Basal Segment	L/2=λ	50.05	0.5
Length of Linguoid	f	73.36	0.732867
Altitude of Basal Segment	g	13.55	0.135365
Lingoid Altitude	h	86.91	0.868232

Table 6.1
The Parameters of the Unitary Linguoid

And draw these dimensions on the Linguoid so:-

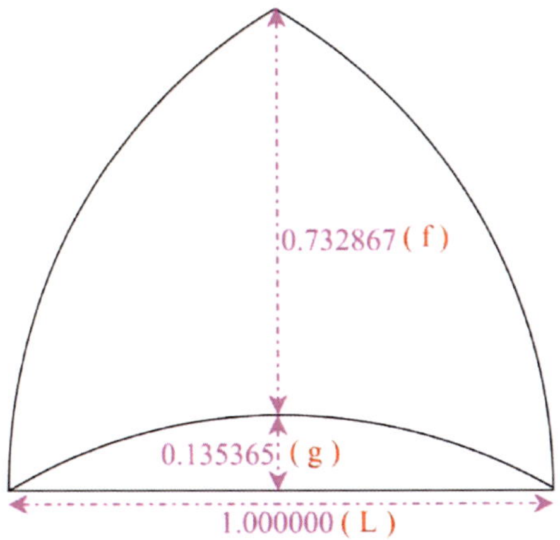

Figure 6.16
Annotated Linguoid

We next check that the flanks of the Linguoid *approximate* a parabola by fitting a red dashed trace according to the following equation:-

$$y = \alpha x^2 + h$$
$$= -3.472928x^2 + 0.868232h$$

Equation 6.2

where the Altitude h is f+g.
This gives us:-

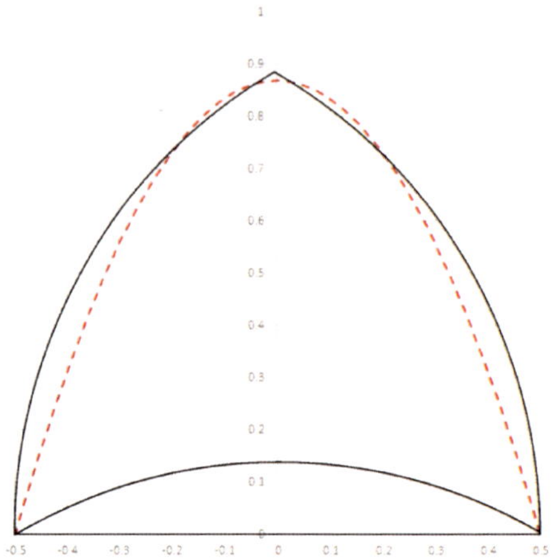

Figure 6.17
A Unitary Linguoid Approximates a Stable Parabola

Next we can fit a reinforcing circle within the linguoid, and nest within that a secondary trefoil, if desired:-

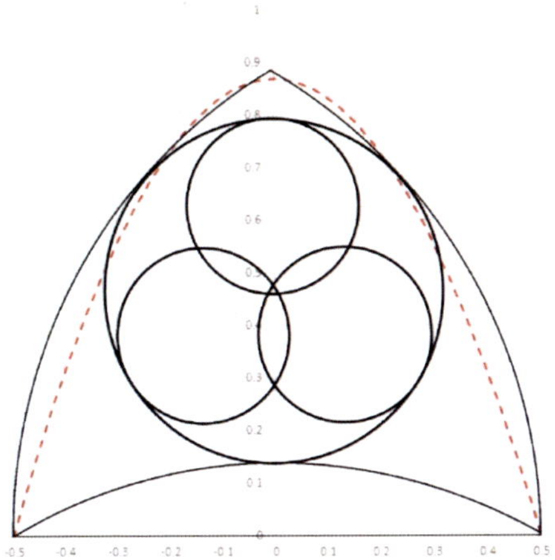

Figure 6.18
A Circle and Trefoil within the Primary Trefoil's Linguoid

The interior circles now give a plan for lead kames to support colored glass that (for illustration) we may install in this way:-

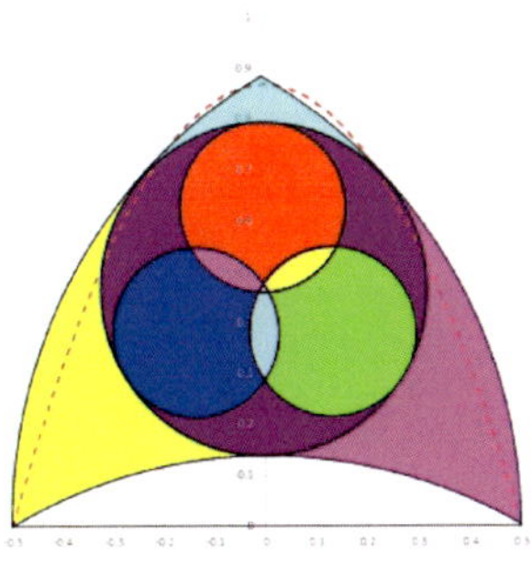

Figure 6.19
Colored Linguoid

This plan may now be replicated and scaled to draft a Gothic-style stained-glass lancet window with three tall lights:-

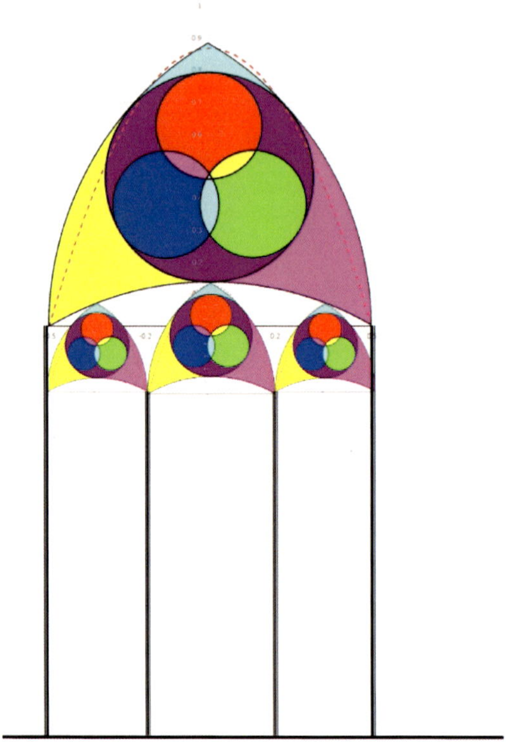

Figure 6.20
A Three-Light Lancet Window

Lichfield Cathedral

Lichfield is an affluent and well-maintained city of about 35000 inhabitants in Central England. Its co-ordinates are SK115097: 52.682°N 1.829°W. The siting of Lichfield may have been influenced by the Roman settlement of Letocetum, a small Roman staging town on Watling Street, about two miles to the South. The modern town was however developed at its current site by St Chad, a major evangelist of the Mercians, in the year 669AD.

Lichfield has a major medieval cathedral whose current fabric dates to its building in 1195AD to 1249AD, with restorations to repair damage sustained during the English Civil War (1642 – 1651) and structural repairs by the Victorians, especially George Gilbert Scott (1811 – 1878).

The walls of the great church bulge out slightly due to the weight of the stone overcroft. The walls are buttressed both with engaged piers and flying buttresses.

It is one of England's great Gothic cathedrals, unique in Britain for its three tall spires between 58 and 77 meters high, and visible for many miles.

The Anglican church is the seat of a bishop since the seventh century and is dedicated to St Chad and the Mother of Christ St Mary.

Lichfield Cathedral is veritably haunted by trefoils, both as blind decorations and numerous stained-glass windows in the clerestory and the lancet windows of The Lady Chapel (the Lady Choir in earlier texts). Figure 6.21 is a photograph taken on the morning of 8 November 2025 depicting the Eastern exterior of the Lady Chapel:-

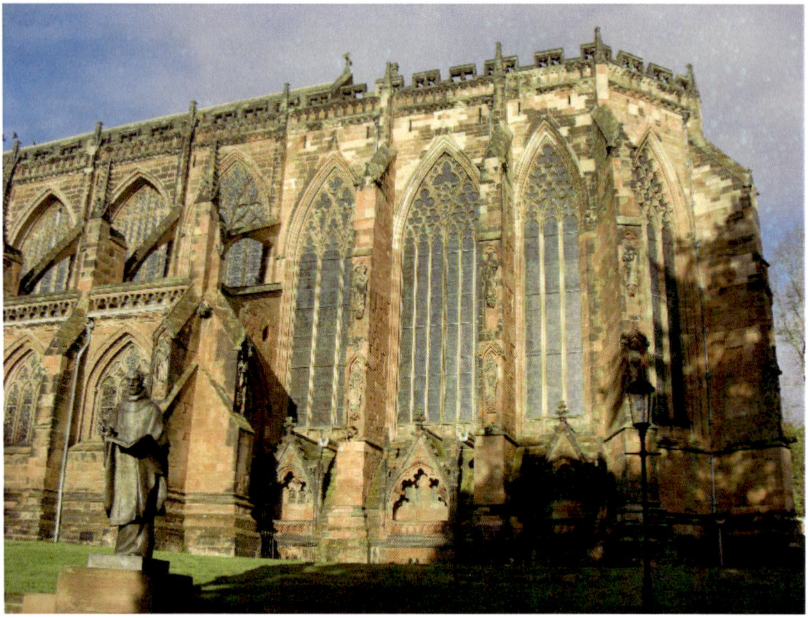

Figure 6.21
The Lady Chapel of Lichfield Cathedral
From the East

You can see four of the glorious lancet windows on the Lady Chapel with up to ten trefoils in their cuspate arches. The trefoils are often ornate and diverse in size. The most prominent display is around (0.55, 0.75). Trefoils also grace the heads of the lower nave windows exampled about (0.05, 0.45).

Note the flying buttress about (0.2, 0.6).

Figure 6.22 shows the South Wall in a more westerly part of the cathedral where numerous trefoils grace the heads of the lower lancet windows, and three trefoils are set within Rouleaux triangles along the clerestory.

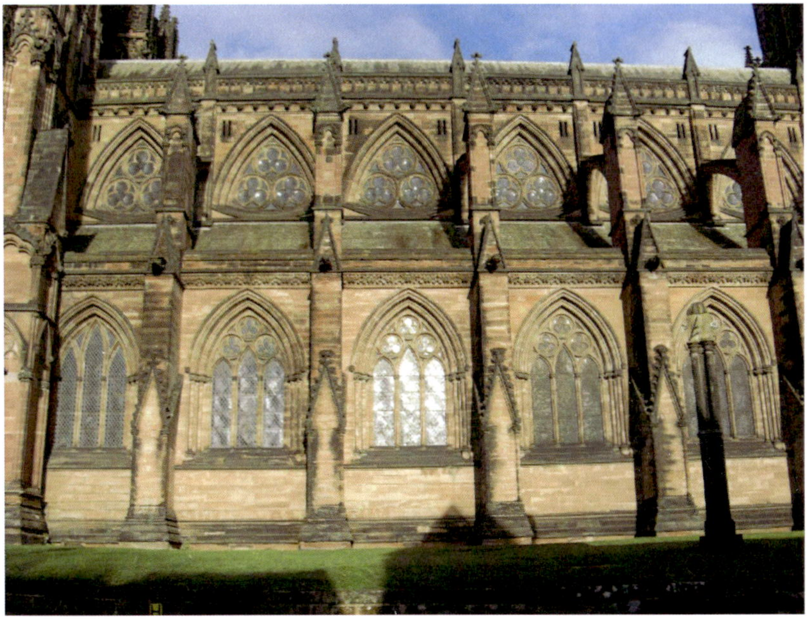

Figure 6.22
Trefoils and Rouleaux Triangles along Lichfield Cathedral's South Wall

Inside the structure, a sunny day allows you to enjoy the contained stained glass in all its magnificence.

Figure 6.23 is a scene taken from the western threshold of the Lady Chapel. The North Wall has trefoils at the top of the plain clerestory windows, and also ornate trefoils in a lower friese. Those north wall decorations are depicted in Figure 6.24.

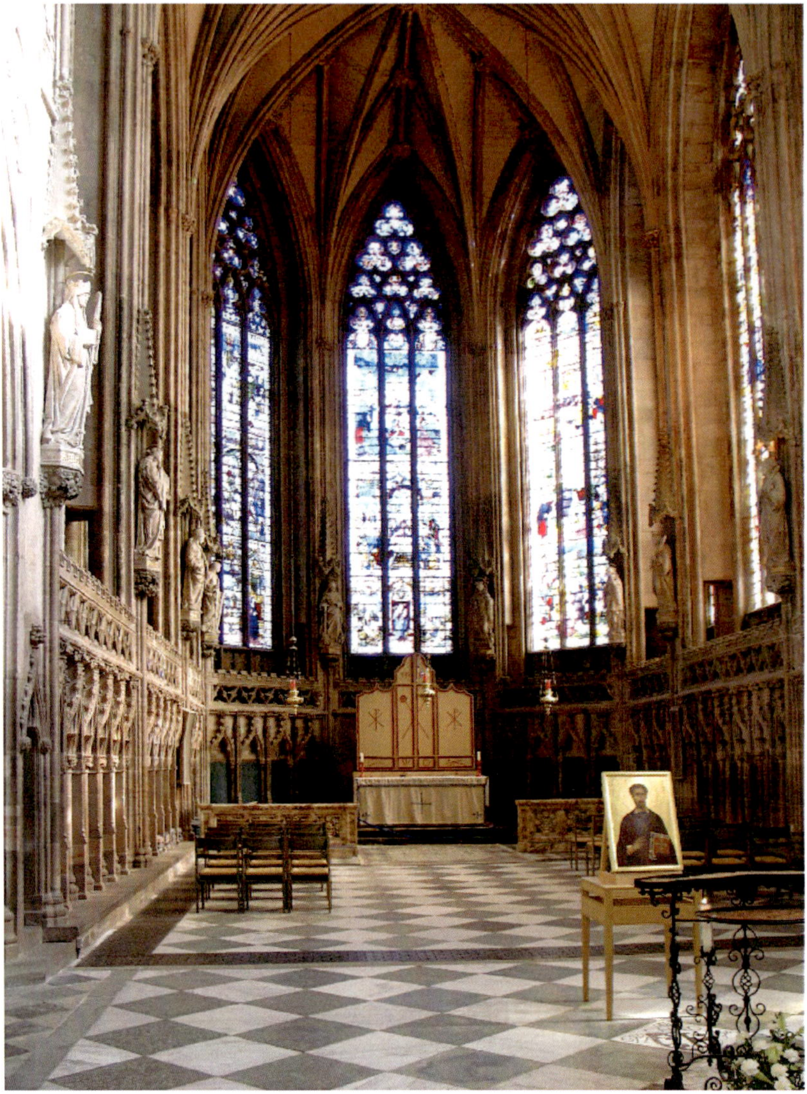

Figure 6.23
Inside the Lady Chapel

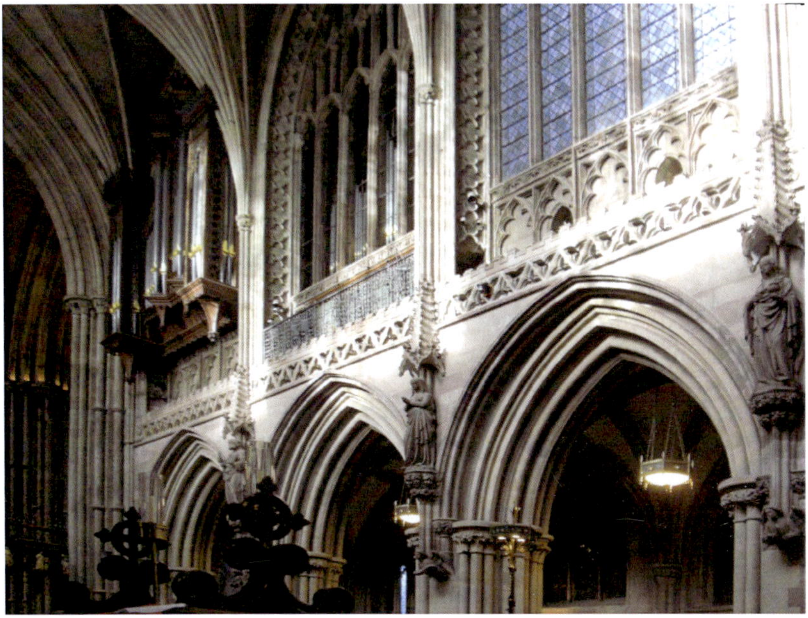

Figure 6.24
Trefoils on the North Wall of Lichfield Cathedral

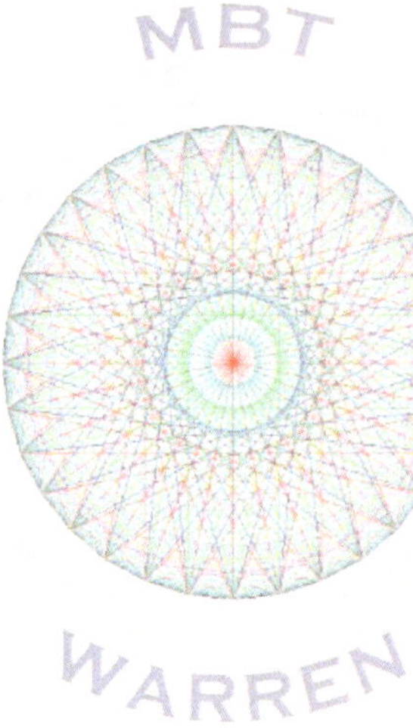
MBT
WARREN

CHAPTER SEVEN
SCIENCE

Physics:- Dimensional Analysis

At the fundament of mathematical physics lies the concept of dimension, a very powerful thought tool for identifying tenable relations in physics and assisting the development of formulae that articulate physical parameters.

For example, we know that Force is the product of Mass times Acceleration, so that Weight, a force, is the Mass of an object multiplied by the local Acceleration due to gravity that varies a little from place to place but for the Earth's Surface is always near to 9.80665 meters per second-squared.

An Acceleration is always measured in meters per second per second because it is a change in Velocity per unit Time.

So in general, independent of our system of units, Acceleration is length×(time unit)$^{-2}$, or to put it concisely LT^{-2} whilst the Velocity to which it is closely akin is measured in meters/second or more generally $L/T = LT^{-1}$.

Velocity and Acceleration are examples of Kinematic parameters because in the system of Mechanical Dimensions MLT they are independent of Mass M. When physical parameters involve Mass, as does weight or any other force, then they are termed Dynamic parameters. So:-

$$F = ma = MLT^{-2}$$

Equation 7.1

where F is Force; m is Mass; and a is Acceleration.

We are taught that Energy, E, is Force (MLT^{-2}) acting through a certain Distance, d, which has the dimension L, so that we may dimensionalise Energy as:-

$$E = Fd = mad = M.LT^{-2}.L = ML^2T^{-2}$$

Equation 7.2

Table 7.1 lists the dimensions of several common parameters in Mechanics, the department of Physics which deals with considerations of kinetics, force and energy.

There is nothing supernatural about dimensional analysis: It is just a device to assist us to define how things are measured and how they relate.

Figure 7.1 is a basic trefoil relation of Mass (M), Length (L) and Time (T) is terms of the association of mechanical parameters, whilst Figure 7.2 develops the picture with the named parameters.

Figure 7.1 hardly provides legible accommodation for all our parameters, and it is better to load "symbols only", as in Figure 7.2 which allows us sufficient space to include Dynamic Viscosity, μ.

Place time, T, in the top lobe of your trefoil, because Time is the indispensable attribute of all mortal things including whole universes, since spiritual things are independent of time. That is the difference between Physics and Metaphysics.

Physical Agent or Entity	Symbol	M	L	T
Mass	m	1		
Length	l		1	
Time	t			1
Area	A		2	
Volume	V		3	
Density	ρ	1	-3	
Velocity	v		1	-1
Acceleration	a		1	-2
Frequency	f			-1
Period	1/f			-1
Wavelength	λ		1	
Force	F	1	1	-2
Pressure	p	1	-1	-2
Power	P	1	2	-3
Energy	E	1	2	-2
Surface Tension	γ	1		-2
Kinematic Viscosity	ν		2	-1
Dynamic Viscosity	μ	1	-1	-1

Table 7.1
Mechanical Dimensions of Some Useful Parameters

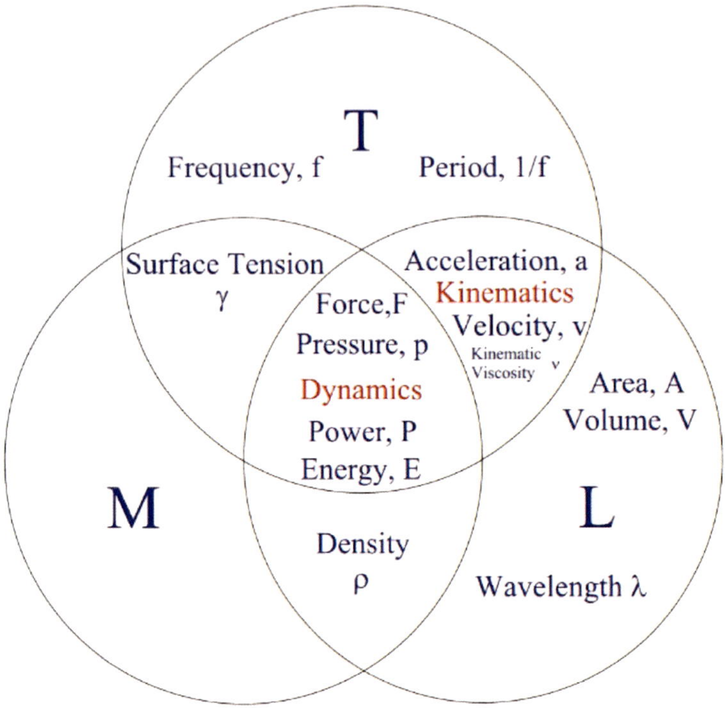

Figure 7.1
The Association of Physical Parameters
In
Mechanics

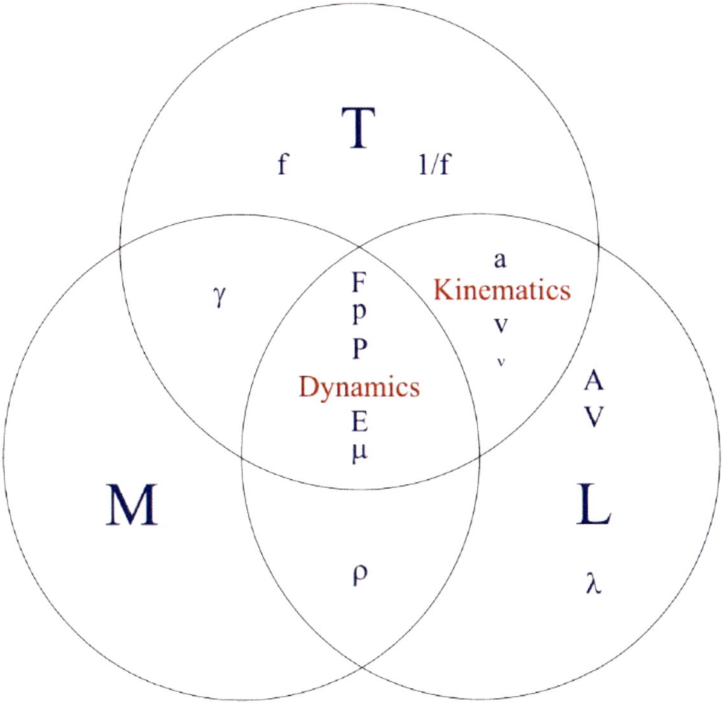

Figure 7.2
The Association of Physical Parameters
In
Mechanics
(Symbols Only)

Glucose Chemistry

Glucose is a colorless, sweet-tasting, odorless, crystalline substance, hygroscopic and soluble in water in most concentrations. It is comprised of Carbon (C) usually a black refractory pseudo-amorphous but crypto-crystalline substance (graphite) or rarely a colorless hard obdurate crystal (diamond); Hydrogen, a very light fugacious colorless gas at 20°C, flammable in Oxygen; and Oxygen a denser, also colorless and odorless, highly reactive gas.

The overall chemical chemical formula of Glucose, which belies a complex molecular structure, is $C_6H_{12}O_6$.

Glucose occurs in nature, admixed with Fructose (also $C_{12}H_{12}O_6$, but with a different molecular structure) in cane sugar. But it is ubiquitous in vegetation, at one concentration or another.

Glucose is essential to almost all known life-forms because it is metabolized to yield energy to power the organism. The metabolization of Glucose is call glycolysis.

Without Glucose death supervenes.

Glucose is one of many thousands of known "carbon chemicals", formed by the covalent-capture of many other elements, especially related non-metals, and most especially Hydrogen and Oxygen.

Many of these carbon-based chemicals are, like glucose, Carbohydrates involving carbon, hydrogen and oxygen in their molecules; or else Hydrocarbons such as the closed Kekule ring flammable oily liquid Benzene (C_6H_6) or the open-chain Octane (a gasolene) (C_8H_{10}).

But chemicals involving Carbon and Oxygen only are also viable and Carbon Dioxide (CO_2) is one of the ultimate waste products of glucose metabolism as is Dihydrogen Oxide (Water) (H_2O). Lactic Acid ($C_3H_6O_3$), a sour-smelling liquid, is also a waste product of glucose oxidation, and in fasting animals or diabetics also Acetone (C_3H_6O), a limpid, florally-smelling, highly evaporative colorless liquid. Yeasts can also excrete Ethanol (C_2H_5OH), which is common alcohol, also a limpid, evaporative, florally-smelling liquid which also looks like water.

Most carbon chemicals are of the first industrial and pharmacological usefulness, as well as essential to life.

Table 7.2 lists selected basic carbon-chemicals and their associated reagents.

The simpler molecules tend to represent waste metabolic end-products such as water (H_2O) and carbon dioxide (CO_2) in addition to Glycerol ($C_3H_8O_3$), Lactic Acid ($C_3H_6O_3$) and Methane (CH_4).

Phosphine (PH_3) is a colorless gas which burns spontaneously in air when facilitated by the presence of Diphosphane (P_2H_4), a colorless liquid. Because plant and animal remains contain Phosphorus (P) the decay of organic remains in for example graveyards and swamps leads to clouds of dilute phosphine in still airs, and at night its slow, cold combustion produces a feeble lurid light often mistaken for supernatural manifestations. In higher concentration both phosphine and diphosphane are poisonous. Phosphine should not be confused with Phosgene ($COCl_2$), a highly corrosive and toxic poison gas used in the First World War.

Phosphorus is essential to Glycolysis, the exothermic metabolization of Glucose.

The introduction of a fourth element indicates that tripartite structures such as the Trefoil are inadequate to the representation of organic chemical inter-relations and that we must develop quatrefoils or five-or-six lobed figures if we mean to represent the associations of glycolysis.

For “Some Simple Covalent Compounds” involving C, H and O only, this trefoil may be adequate as in Figure 7.3.

Figure 7.4 illustrates the progress of glycolysis for the conversion of glucose into pyruvate to release energy for the sustentation and development of the organism. As stated, the trefoil is inadequate at this level of detail. To cover C, H, O and P, (a “CHOP” system) we need a quatrefoil such as Figure 7.5, and for the System C, H, O, N, P we would need a five lobed structure such as Figure 7.6

Agent or Entity	Formula	C	H	O	N	P	Cl
Some Carbon-Chemistry Elements							
Carbon	C	1					
Hydrogen	H		1				
Oyygen	O			1			
Nitrogen	N				1		
Phosphorus	P					1	
Chlorine	Cl						1
Some Simple Covalent Compounds							
Carbon Dioxide	CO_2	1		2			
Carbon Monoxide	CO	1		1			
Water	H_2O		1	1			
Hydrogen Peroxide	H_2O_2		2	3			
Benzene	C_6H_6	6	6				
Octane	C_8H_{10}	8	10				
Methane	CH_4	1	4				
Acetylene	C_2H_2	2	2				
Glucose	$C_6H_{12}O_6$	6	12	6			
Fructose	$C_6H_{12}O_6$	6	12	6			
Phosgene	$COCL_3$	1		1			3
Phosphine	PH_3		3			1	
Diphosphane	P_2H_4		4			2	
Intermediate Glycolysis Products							
Glycerol	$C_3H_8O_3$	3	8	3			
Lactic Acid	$C_3H_6O_3$	3	6	3			
Acetone	C_3H_6O	3	6	1			
Ethanol	C_2H_6O	2	6	1			
Phenylalanine	$C_9H_{11}NO_2$	9	11	2	1		
Adenosine Triphosphate	$C_{10}H_{16}N_5O_{13}P_3$	10	16	13	5	3	
Key Glycolytic Pathway Chemicals							
Glucose	$C_6H_{12}O_6$	6	12	6			
Glucose-6-Phosphate	$C_6H_{13}O_9P$	6	13	9		1	
Fructose	$C_6H_{12}O_6$	6	12	6			
Fructise Phosphate	$C_6H_{13}O_9P$	6	13	9		1	
Pyruvate	$C_3H_3O_3$	3	3	3			

Table 7.2
Carbon Chemicals and
Some Associated Reagents

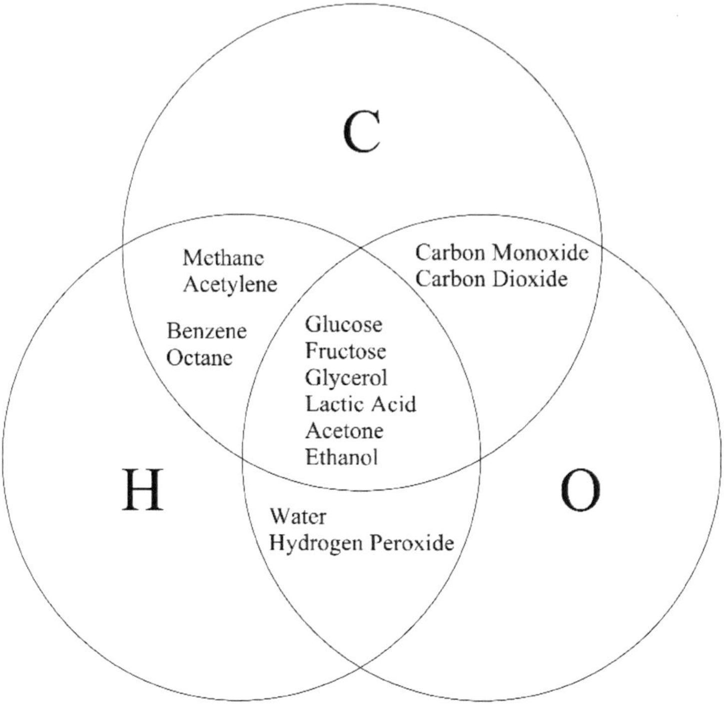

Figure 7.3
The Associations of
Some Simple Covalent Compounds

Catalysed Reactions involving ATP and Mg^{++}

Glucose
$C_6H_{12}O_6$
↓
Glucose 6-phosphate
$C_6H_{11}O_6..PO_3^{-2}$
↓
Fructose 6-phospahte
$C_6H_{11}O_6..PO_3^{-2}$ (isomer)
↓
Fructose 1,6-biphosphate
$C_6H_{10}O_6..2(PO_3^{2})$
↓
Glyceraldehyde 3-phosphate and
Dihydrooxyacetone phosphate
$2\times(C_3H_5O_3.PO_3^{-2})$ (isomeric twins)

Reversible Reactions involving ATP and Mg^{++}

1,3-biphosphoglycerate
$C_3H_4O_4.2(PO_3)^{-2}$
↓
3-phosphoglycerate
$C_3H_4O_4.PO_3^{-2}$
↓
2-phophoglycerate
$2\times C_3H_5O_4.PO_3^{-2}$
↓
Phosphoenolpyruvate
$C_3H_3O_3.PO_3^{-2}$
↓
Pyruvate
$C_3H_4O_3$

Figure 7.4
The Progress of Glycolysis

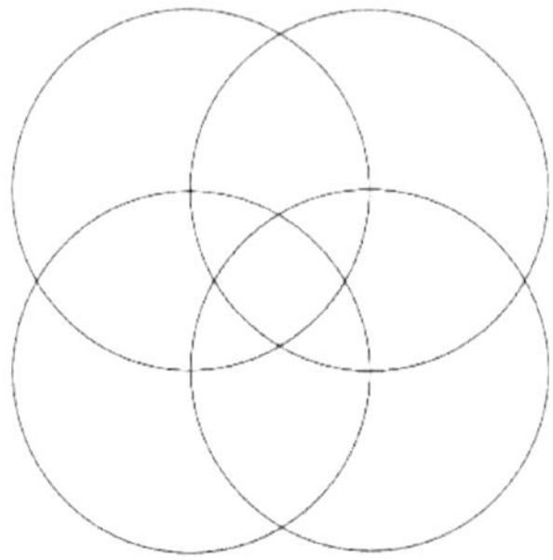

Figure 7.5
A Quatrefoil

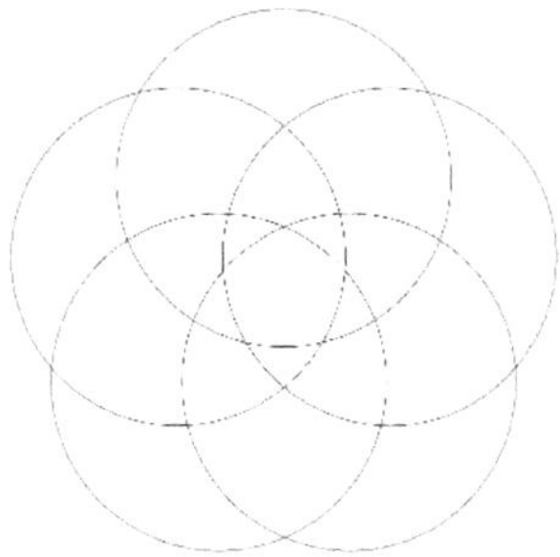

Figure 7.6
A Five -Lobed Figure

In old-fashioned terms the skin of the Earth is Continental Crust which floats upon a layer of Basic (i.e. non-acidic) abyssal "Basement" rock which slowly convects from the deepest regions of the planet, forming the outer Crust which it moves and molds.

The Crustal rocks are not only continuously augmented by basement but also continuously re-cycled as they are riven and reformed by the intestinal workings of the Earth, ultimately driven by radiogenic heat from natural nuclear decay as well as what remains of the primal heat of gravitational aggregation.

If we focus our attention upon the Crustal rocks of the great land masses, including Britain as it sits upon the Eurasian Continental Plate, the Crustal Rocks fall into three great facies:-

(A) **Igneous**

These are primal Crustal rocks which form upon or else shallowly within the Earth's Crust by Intrusion into the Crust at a depth of hundreds or thousands of meters; or Extrusion by actual eruption at the surface or submarine of the planet as molten rock.

Intrusive rocks are often exposed by erosion of superincumbent rock and include acidic Granites or allied rocks such as Pegmatite which typically comprise large crystals of

Quartz (SiO_2);
Feldspar ($KAlSi_3O_8$–$NaAlSi_3O_8$–$CaAl_2Si_2O_8$);
and the two Mica minerals
Muscovite ($KAl_2(AlSi_3O_{10})(F,OH)_2$)
and Biotite ($K(Mg,Fe)_3(AlSi_3O_{10})(F,OH)_2$).

Extrusive rocks outcrop by definition and include the acidic Rhyolite, typical of explosive cone volcanoes; and the more Basic (alkaline) trap extrusives typical of Hawaii, Iceland

and the Scottish Inner Hebrides which are represented by Basalt, a free-flowing rock when molten, which contains such as

Plagioclase ($NaAlSi_3O_8 - CaAl_2Si_2O_8$);
Hornblende
($(Ca,Na)_{2-3}(Mg,Fe,Al)_5(Al,Si)_8O_{22}(OH,F)_2$);
and Olivine ($(Mg,Fe)_2SiO_4$).

(B) **Metamorphic**

These are rocks altered by enormous heat and or pressure from the Earth's slow convulsions. They are formed from the re-crystallisation of Igneous rocks, as we discussed above, or Sedimentary rocks which we shall discuss below.

Slates and Schists are typical of pressure-changed "Regional" metamorphics; and Marbles typical of heat-changed "Contact" metamorphic rocks.

Slates are metamorphosed mudstone with hydrous aluminum phyllosilicate mineralogies (e.g. Kaolin, $Al_2Si_2O_5(OH)_4$) and Schists of similar mineralogy to Granite but often with more Mica.

Marble is simply baked Limestone ($CaCO_3$) or Dolomite ($CaMg(CO_3)_2$), but usually with admixed vein minerals.

(C) **Sedimentary**

Sedimentary rocks are stratified depositional indurated sediments such as Sandstone (SiO_2) formed from sand; Shale and Mudstone formed from Alumino-Silicates derived from such as Feldspar; and Limestone ($CaCO_3$).

Sandstone and Shale are Clastics whose sediments were formed by the high-energy mechanical breakage, weathering and grinding of Igneous or Metamorphic crystals; whilst most primary Limestones ($CaCO_3$) originated as low-energy shallow-water accumulations of the skeletons of marine microorganisms.

Within the Sedimentary group of crustal rocks we may identify the silicious Arenites (sandstones); the Aluminous Argillites (shale and mudstone); and the Calcareous Carbonates (limestones).

There are myriads of different mixtures and varieties of all these gross rocks.

The Sedimentary rocks are illustrated in the Trefoil of Figure 7.7:-

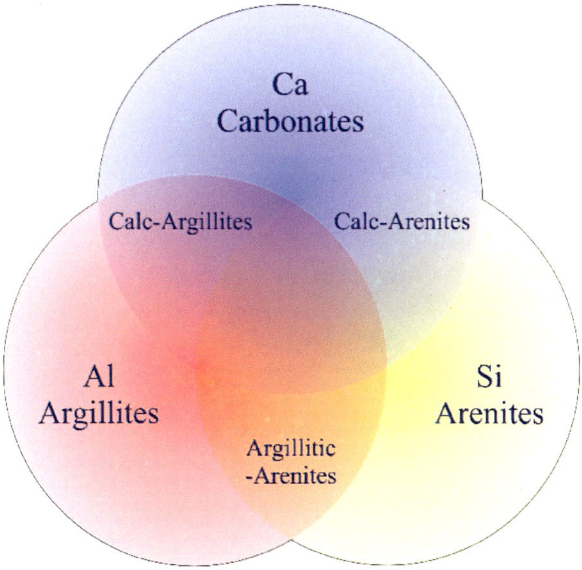

Figure 7.7
The Sedimentary Trefoil

Agriculture

The key to prosperous crop production and sustained healthy yields is the fertilisation of the land.

Plants convert simple chemicals into complex ones upon which humans and other animals can feed, act and grow.

The three essential elements are:-

(A) **Nitrogen (N)**

This gas occurs in the air, and indeed legumes like peas and beans can "fix" atmospheric nitrogen with the help of microorganisms who live symbiotically in the legumes' roots.

Other plants need nitrogen which is dissolved in water as solutions of chemicals, which chemicals are typically solid when dry. Examples of such soluble chemical fertilisers are Ammonium Nitrate ($NH_4.NO_3$) and Ammonium Sulfate ($(NH_4)_2.SO_4$).

All plants benefit from nitrogen but it is especially good for cereals, brassicas and leafy greens, as well as root-crops such as carrots, radishes and onions. It is also good for fruiting vegetables like tomatoes and cucumbers, as well as oilseed rape and grass.

Nitrogen aids chlorophyll development and protein synthesis.

(B) **Phosphorus (P)**

Phosphorus is a solid non-metallic element but in combination with metals or ammonium forms soluble colorless crystals which dissolve in water. These dissolved phosphates are useful to plants, being absorbed by their roots with water.

Phosphates are essential for all or most plants, but are especially beneficial for legumes, grains, root crops like potatoes, and commercial blooming flowers.

Phosphates stimulate root development and ATP production for cell metabolism, whilst also increasing disease resistance.

(C) **Calcium (Ca)**

In its pure form Calcium is a soft, white metal that reacts strongly with water to evolve hydrogen and leave Calcium Hydroxide ($Ca(OH)_2$). It is this and other barely-soluble salts of Calcium, in particular Calcium Oxide (CaO) which reacts with water to make Slaked Lime (Calcium Hydroxide, $Ca(OH)_2$); and also Calcium Carbonate ($CaCO_3$), as Limestone a natural rock. Lime and Limestone are especially interesting to the farmer as vectors of bioactive calcium.

Calcium is an alkali metal and reacts with acids to produce neutral salts like the hydroxide and the carbonate.

The upland areas of Britain are covered with blanket peat, the wet semi-decayed detritus of heathers, bilberries, lings and other acid-loving mountain plants. Peat and heavy clays require dressing with lime to become chemically-neutral and agriculturally productive. Lime dressing comminutes such soils, rendering them drier, more workable and chemically more yielding.

The North and West uplands of Britain feature many derelict limekilns where plentiful limestone was roasted with coal to produce Quicklime (CaO) and Carbon Dioxide (

CO_2) which was a gas wasted to the atmosphere. In a secondary rection with water, the quicklime was spread neat upon the fields if it was desired to suppress weeds, or else "slaked" with water to produce the hydroxide of calcium, which was then spread as a manure. This enabled such land to be unlocked for agriculture, especially the potato.

We may represent Ca-N-P as a triune system in the Trefoil, together with representative fertilisers and soil conditioners:-

Species	Chemical Formula	Nominal Color
Ammonium Nitrate (N)	$NH_4.NO_3$	Red
Bone Ash (Ca, P)	$Ca_5(PO_4)_3(OH)$	Green
Calcium Ammonium Nitrate (CAN) (Ca, N)	$5Ca(NO_3)_2.NH_4NO_3$	Magenta
TriAmmonium Phosphate (N, P)	$(NH_4)_3PO_4$	Orange
Apatite (P)	$Ca_5(PO_4)_3(F,Cl,OH)$	Green
Superphosphate	$Ca(H_2PO_4)_2$	Green
Lime (Ca)	CaO	Blue

Table 7.3
The Fertiliser System (Ca-N-P): Chemical Formulae

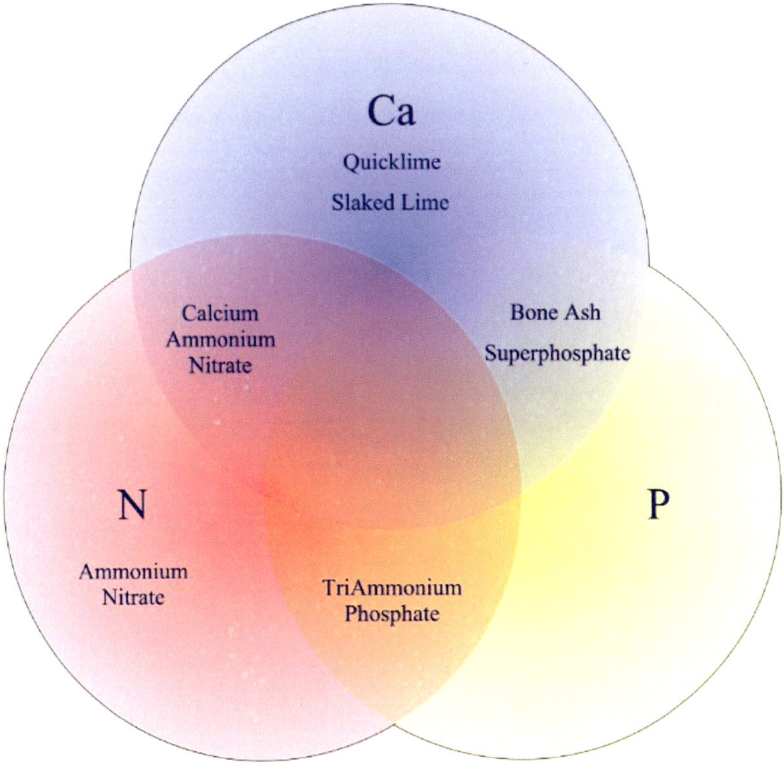

Figure 7.8
The Fertiliser System (Ca-N-P)

Bone Meal is essentially a crude mixture of residual proteins and bone ash, and would sit in the Rouleaux Triangle middle of the Trefoil. It is hardly a vegan supplement, or a cost-efficient one, but is chemically possibly as close to an optimal dressing as is practicable. Otherwise, scientifically-optimised blends of Superphosphate, TriAmmonium Phosphate and CAN are probably best tailored to your crop.

Mineralogy

The Garnet System

Garnet is a characteristic late-crystallising product of regional metamorphism produced by the high-pressure, high temperature baking of argillites (pelites).

I well remember during the September of 1974 mistakenly mapping a landscape of Dalradian Garnetiferous Schist at the commencement of my undergraduate mapping project on the Northern flanks of Schiehallion, a shapely and famous Scottish mountain, against which Nevill Maskelyne weighed the Earth in 1774. That part of the Highlands was a very bleak slope of grey, course shelves of rock interspersed with clumps of stunted birch and ash trees. I had seen no living creature for days and I started talking to myself. The garnets were up to a centimeter in diameter, ubiquitous and very red, but mottled and opaque: Not gem quality be any means.

I was rescued by the late Dr Jack Edward Treagus, who set me to work mapping the Blair Atholl Series, a complex and contorted assemblage of altered argillites and marbles, at its Lochan an Daim emergence, in a shallow valley between those garnetiferous schists and the massive Schiehallion Quartzite to the South.

Despite the early problems, it was one of the best experiences of my life.

The General Formula of the Garnet is $X_3Y_2.(SiO_4)_3$, where X and Y are metallic atoms. Table 7.4 presents the Garnets with their Nominal Colors:-

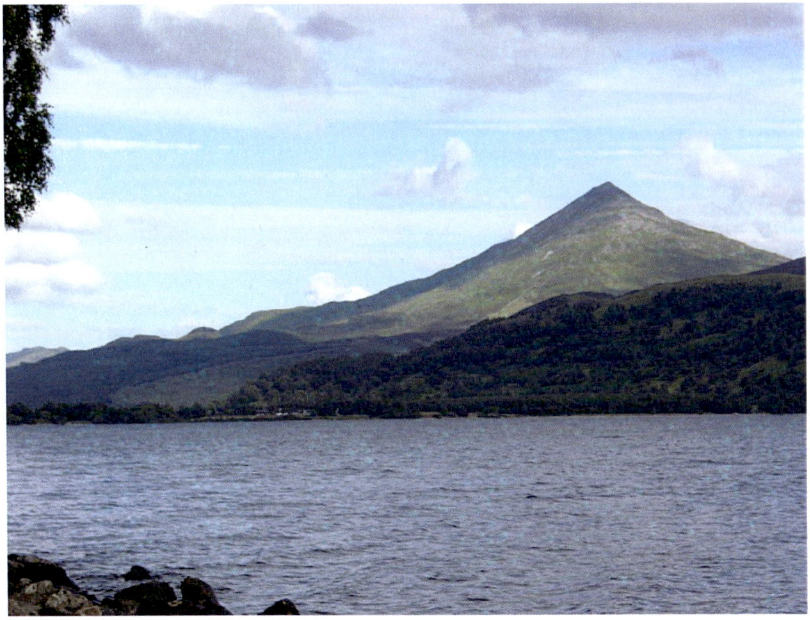

Figure 7.9
Schiehallion Mountain from Loch Rannoch

Species	Chemical Formula	Nominal Color
Pyrope	$Mg_3Al_2Si_3O_{12}$	Red
Almandine	$Fe^{2+}{}_3Al_2Si_3O_{12}$	Purple
Uvarovite-Grossularite	$Ca_3Cr_2Si_3O_{12}$	Green

Table 7.4
The Garnet Gemstones

Gold and garnet jewelry has characterised male dress in the West since at least Classical times. Much of the Staffordshire

Hoard and other ancient treasure troves comprises such, and my Late Father's wedding band was (and remains) a thick ring of eighteen-carat (75%) gold set with three fine blood-red pyrope garnets.

Using a Trefoil we may represent the Garnets and their characteristic colors in these terms:-

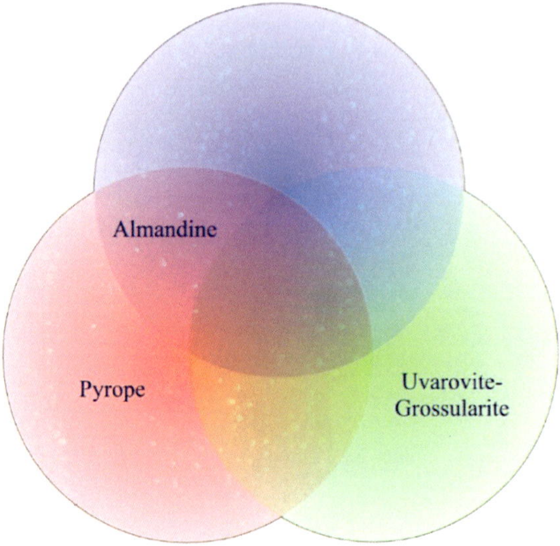

Figure 7.10
The Garnet Trefoil

The Corundum System

Corundum is a very hard mineral crystal which is only rarely transparent or of gem quality. Opaque, black-looking corundum is much more common as Emery, which is used for industrial grinding processes being much cheaper than the even harder diamond, an allotropy of Carbon.

Corundum occurs in a wide variety of igneous and metamorphic rocks.

Gem quality red Rubies are still mined in Burma (Myanmar) and Ceylon (Sri Lanka) was once famous for its superb blue Sapphires.

Notwithstanding, almost all gem-quality rubies and sapphires are today made artificially and whilst very expensive are cheaper than natural stones.

Note that Emerald is *not* Green Sapphire: Emerald is a Beryl, Beryllium Aluminum Silicate, with the chemical formula $Be_3Al_2(SiO_3)_6$.

The General Formula of Corundum is $(X,Y)Al_2O_3$ where X and Y are metallic atoms. Table 7.5 presents the gem-grade Corundums with their Nominal Colors:-

Species	Chemical Formula	Nominal Color
Ruby	$(Cr)Al_2O_3$	Red
Green Sapphire	$(Fe)Al_2O_3$	Green
Yellow Sapphire	$(Fe,Ti,Cr)Al_2O_3$	Yellow
Sapphire	$(Fe,Ti)Al_2O_3$	Blue

Table 7.5
The Corundum Gemstones

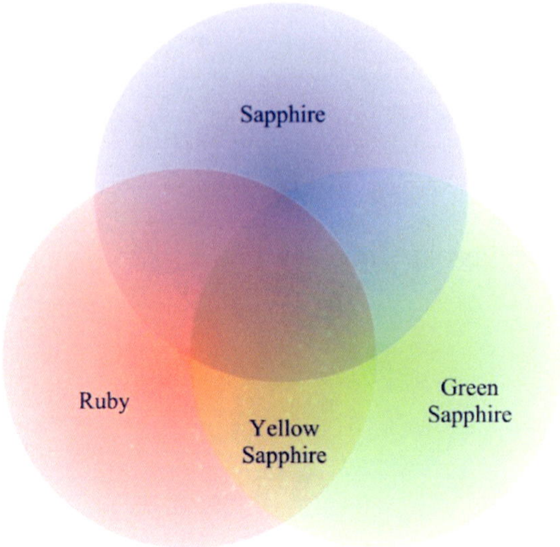

Figure 7.11
The Corundum Trefoil

Metallurgy

Metallurgy is the scientific study of metallic elements and their physical and chemical relations, especially their alloys the creation and characterisation of which constitutes an immense technology to which I cannot do justice.

I will present some Trefoils to illustrate some of the major metal alloy systems.

Table 7.6 summarises the chemistry and nominal colors of several significant metallic alloys, which may themselves be simple mixtures or crystalline chemical compounds by atomic lattice substitutions.

It is very clear that the Trefoil is sub-optimal for representing alloys with four, five or six constituent elements but in the case of, for example, the aviation alloy Aluminum 7075 which has six constituents we may arrange that each of the three lobes contains a pre-ponderating element: Al (89.97%); Zn (5.6%); Mg (2.5%): And the three linguoids minor constituents: Cu (1.4%); Mn (0.3%) and Cr (0.23%).

During the Second World War my Late Mother was conscripted to work in chemical laboratories because of her exceptional scientific abilities and spent part of her Service developing aviation duralumin in The James Booth Aluminium Company at Kitt's Green, Birmingham, England. She was not allowed to join the Royal Navy, which was her wish, but did so after the War ended.

In such a scheme for Duralumin the basic System is Al-Zn-Mg as shown in Figure 7.12.

The illustrations are more clear-cut with the Copper Alloys Bronze and Brass as shown in Figure 7.13 and more decided still for jewelers' Colored Gold as presented in Figure 7.14.

Table 7.7 lists typical goldsmiths' formulations of varieties of ornamental gold and silver often optimised for color and durability within the constraints of the law.

Color does of course depend upon subjective perceptions and the state of the light, so that the RGB definitions and the rendering in Figure 7.14 should not be taken too literally.

Notably, White Gold is *not* a blend of Copper, Silver and Gold, or at least not usually. White Gold is usually an alloy of Nickel and Gold except in contexts where the wearer is likely to suffer a dermatological allergy to Nickel.

Alloy	Species	Nominal Color	Percentage Weight Composition by Element Fe	C	Ni	Cr	Mn	Mo	Al	Mg	Cu	Sn	Zn	Au	Ag	As	Total
STAINLESS STEEL	Grade-316 Marine	White	69.5		9	19		2.5									100
DURALUMIN	Aluminum 7075	White				0.23	0.3		89.97	2.5	1.4		5.6				100
BRONZE	Bell Metal	Russet									77.5	22.5					100
BRASS	Admiralty Brass (UNS C44300)	Yellow									71.01	1.043	27.91	0	0	0.04	100
GUNMETAL	UK Admiralty (US G C90500)	Yellow									88	10	2				100
YELLOW GOLD	18-karat Gold	Yellow									12.5			75	12.5		100
WHITE GOLD	18-karat White Gold	White			25									75			100
ROSE GOLD	18-karet Rose Gold	Red									23			75	2		100

Table 7.6
Some Important
Metal Alloy Systems

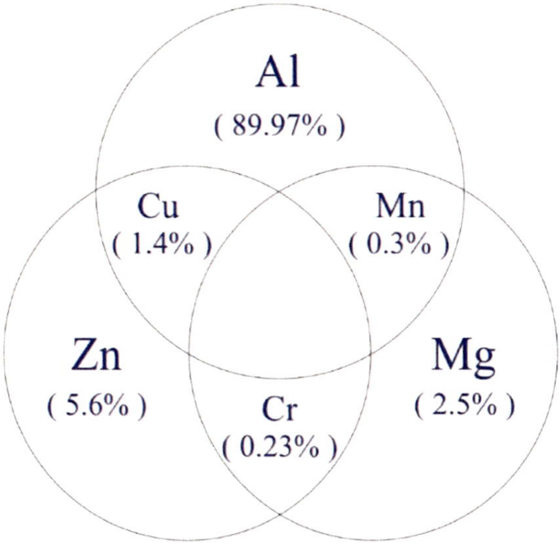

Figure 7.12
The System Al-Zn-Mg
For
DURALUMIN Aluminum 7075

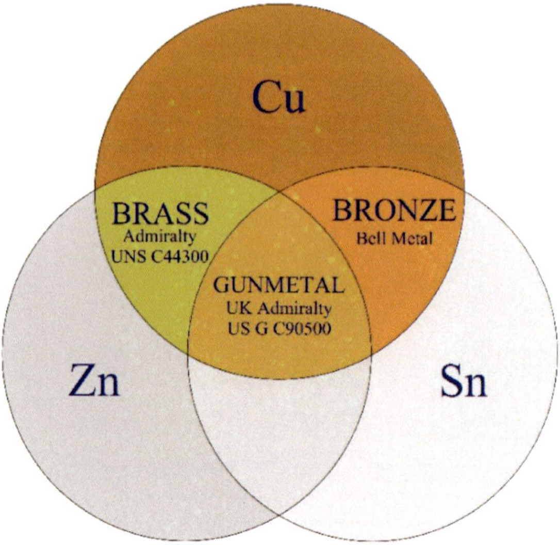

Figure 7.13
The System Cu-Zn-Sn
For
Copper Alloys Bronze and Brass

Alloy	Species	Nominal Color	RGB Formulaic Color			Percentage Weight Composition by Element				
			Red	Green	Blue	Ni	Cu	Au	Ag	Total
PURE GOLD	0.9999 Fine Gold	Yellow	255	215	0			100		100
YELLOW GOLD	18-karat Gold	Yellow	255	215	0		13	75	13	100
WHITE GOLD	18-karat White Gold	White	255	255	244	25		75		100
ROSE GOLD	18-karet Rose Gold	Red	183	110	121		23	75	2	100
PURE SILVER	Silver	White	210	210	210				100	100
STERLING SILVER	Sterling Silver	White	192	192	192		7.5		93	100

Table 7.7
Gold and Silver Alloys

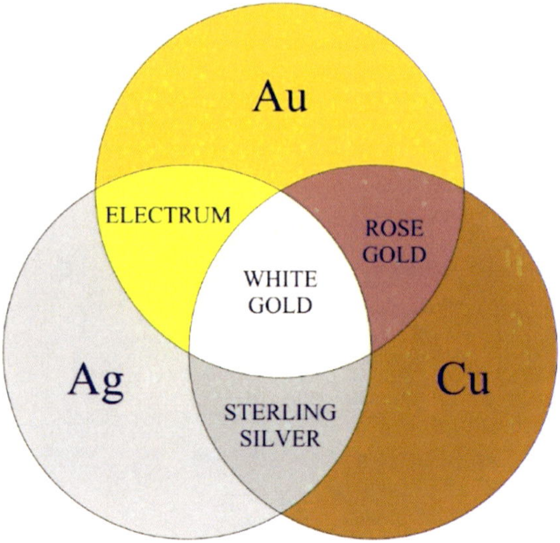

Figure 7.14
The System Au-Ag-Cu
For
Jewelers' Colored Gold

EPILOG

**"Sometimes just the space is what you need.
Not even a solution.
Just the space"**

- Lemn Sissay at 7:28 pm · 5 Jan 2021

A Trefoil is a very simple thing. Not even a solution. Hardly even a space. Just a diagram. An arena for your thoughts, maybe as you begin to organise them.

Consider the humble amphidrome,

Yes, I have never heard of it either, even though I have a Masters in Hydraulics, Hydrology and Coastal Dynamics. It is the hydraulic still point, a tidal node, where two countervailing tides meet on the marine surface.

No, I mean something different here.

I mean a word which contains at least two other words at each end, with a possible third in the middle.

For example, FLEET which resolves to FLEE, to hasten from a hazard; LEET, a Medieval court of law in England; and LEE, the sheltered side of a building or vessel. A FLEET is a collection of ships.

Or FLANK, the exposed side of an advancing army. This resolves to FLAN, a type of pie; LANK, meaning straight and slender, like retted flax or combed blonde hair. The central word LAN is a debased example. It is an abbreviation for Local Area Network, but is nevertheless enunciable.

Such words are *associations* of other words that *connote* other entities that reside, so to say, above and below their plane.

In my idle hours I have identified eleven six-letter English amphidromes like CLOVER; 122 five-letter such as FLEET; and 179 four-letter like BAND. I have no doubt that a competent programmer armed with a word database could abstract many more.

Such amphidromes can be formed, as in the children's game, into word pyramids such as:-

C R R A E N E T H A N K	F L R E A E N T H A N K

And these may be organised into Trefoils like Figure Epi.1 and Figure Epi.2:-

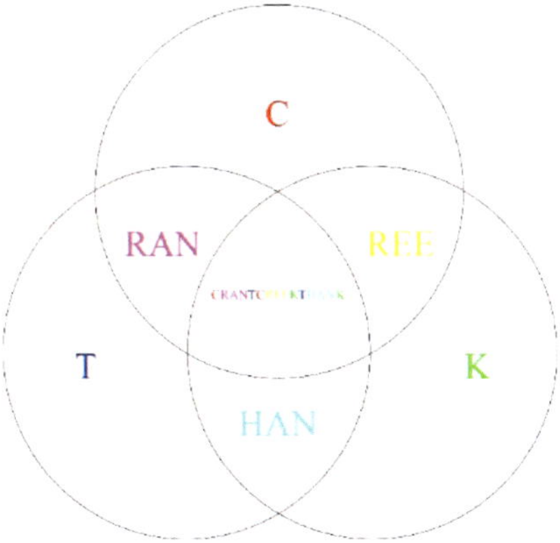

Figure Epi.1
The CRANTCREEKTHANK Trefoil

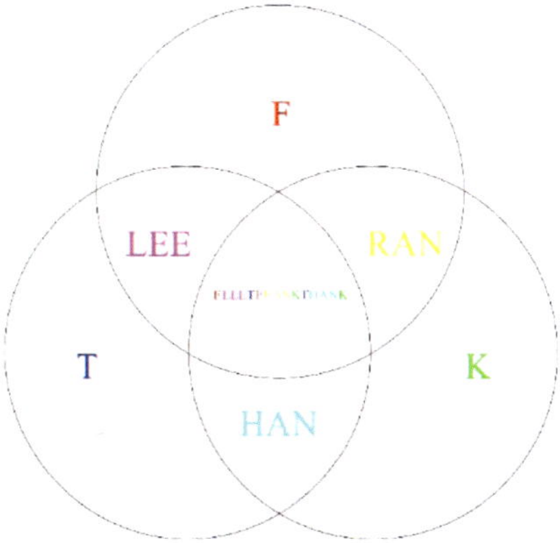

Figure Epi.2
The FLEETFRANKTHANK Trefoil

In our scheme, the central Reuleaux Triangle is assigned a concatenation of three five-letter amphidromes: For CRANT, CREEK and THANK; CRANTCREEKTHANK; and for FLEET, FRANK and THANK; FLEETFRANKTHANK.

We see that we have the makings of a mnemonic passcode because the Permutations, nPr, of three objects chosen *with a given order-of-r* from a collection of 122 five-letter words is:-

$$nPr = 122P3 = \frac{n!}{(n-r)!} = 1771440$$

Equation Epi.1

So that the chance of guessing CRANT.CREEK.THANK *in that order* is one in one million, seven-hundred and seventy-one thousand, four hundred and forty.

It is rather as if (I am British: I unblushingly lard my discourse with "rather as if" and "somewhat", even in mathematical writing. Also I trained as a geologist); it is rather as if the poor Trefoil is sitting meekly upon the plane of his paper is a trackless tract in a wayless world, waiting like any good servant to be addressed.

Albert Einstein solved problems by equivalating Gravity and Inertia on the basis that their perceptible effects were identical. Is this, as a Christian Existentialist might say, Evasion?

Or is this "thinking outside the box"?

There has been a good deal of empirical discovery of phenomena substantiating (not "proving") Einstein's Theory of General Relativity.

Does this have implications for the nature of experience that we would be well advised to investigate?

To quote the late John Lennon, in his 1971 verses "Imagine":-

"Imagine there's no heaven
It's easy if you try
No hell below us
Above us, only sky"

Speaking for myself, I find these questions very far from easy.

What is the Sky? What is the Deep?

I heard a politician the other day say she wished to see "Cranes in the Sky".

Where? What?

Did she mean "crane" as in the lanky North European bird celebrated for nesting atop chimneys? Or the type of cranes deprecated by Trajan which builders use for hoisting materials aloft?

The BBC used to air a programme called "The Sky at Night", about vast emitters of radiation billions of miles away. Is this the same place that cranes inhabit?

Mirriam-Webster defines the sky in these terms:-

"the upper atmosphere or expanse of space that constitutes an apparent great vault or arch over the earth"

This is poetic but seems somehow insipid like warm beer and hardly covers cranes or even jet planes.

Let Wikipedia clarify:-

"The word sky comes from the Old Norse sky, meaning 'cloud, abode of God'. The Norse term is also the source of the Old English scēo, which shares the same Indo-European base as the classical Latin obscūrus, meaning 'obscure'."

In Old English, the term heaven was used to describe the observable expanse above the earth. During the period of Middle English, "heaven" began shifting toward its current, religious meaning.

Reverting to the Anglo-Saxon "Welkin" hardly helps.

There may be no Hell below us, but what is below us?

You may say that you hope it is the floor, especially the floor of the lift when you ride space in Einstein's Elevator.

As a geologist I say a lot is below us, but whatever it is there are only a few thousand kilometers of it before we again re-enter bare and empty Space.

As I say elsewhere, I think in "Congleholme", Heaven and Hell are what you make them.

When I was a student, more than fifty years ago, some superb American hardback geology books were available from Haigh and Hochland's: For a price. They were far superior to anything my own country provided.

Among the sharp and almost photographically-graded pictures were peculiar line diagrams of closely-delineated red and green straight lines. You were supposed to view these through spectacles with one lens green and the other red.

Because I was stereoblind these schemes left me cold.

The intent was that mineral crystals would, as it were, reify as solid objects above and below the plane of the page so that the sizes of inter-planar angles would be manifest, those angles being a constant for any of the finite number of crystal lattices possible for one of the infinite permutations of elemental lattices possible.

Yes, it is confusing and it certainly confused me.

And what of the Deep?

"The Deep" is one of the most romantic of poetic terms, from the depths of thought, a redemption of oblivion, *De Profundis*.

The Deep is the abode of lost gods, prehistoric fish and shipwrecks, all truly "time capsules" from which archaeologists, divers and poets extract The Past.

If The Sky is the Future: Then The Deep is the finite and limitless repository of the Past, Father and Mother to the rocks below us, where we stand in the Sky.

So maybe the Trefoil is like a Tree, with a concentric chronical of rings at its interface with the ground, its branches in the Sky and its roots in the Land, unless it is that sad, ancient Seahenge of a tree set in the tide-washed littoral of Holme, baffled an inversion at the triple point of Land, Sea and Sky.

APPENDIX A
THE NUMERICAL AND SURD EXPRESSIONS OF SELECTED TRIGONOMETRIC VALUES

Angle (degrees)	Angle (Radians)	Values Sine	Values Cosine	Surds Sine	Surds Cosine
15	0.26179939	0.25881905	0.96592583	$\frac{\sqrt{6}-\sqrt{2}}{4}$	$\frac{\sqrt{6}+\sqrt{2}}{4}$
30	0.52359878	0.50000000	0.86602540	$\frac{1}{2}$	$\frac{\sqrt{3}}{2}$
60	1.04719755	0.86602540	0.50000000	$\frac{\sqrt{3}}{2}$	$\frac{1}{2}$
120	2.09439510	0.86602540	-0.50000000	$\frac{\sqrt{3}}{2}$	$-\frac{1}{2}$

APPENDIX B
ZONE DEFINITION GALLERY

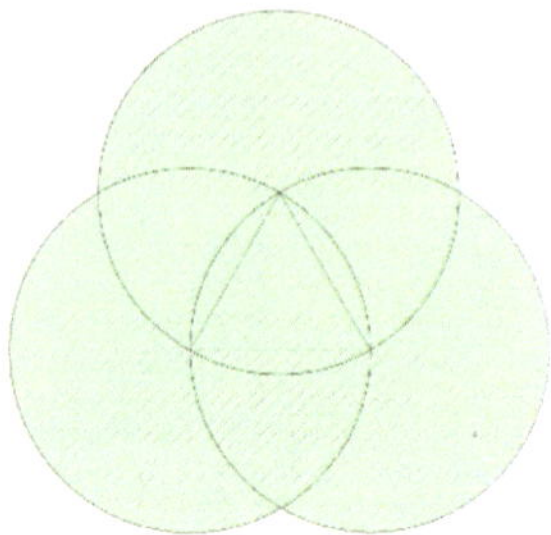

Figure B1
The Whole Trefoil
(The "BLOB")

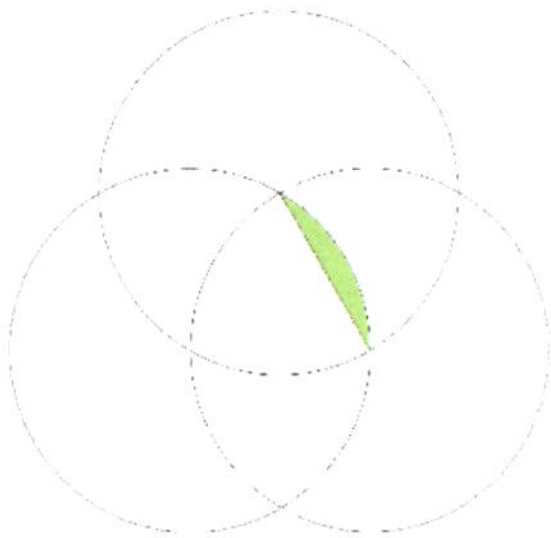

Figure B2
The Minor Segment1

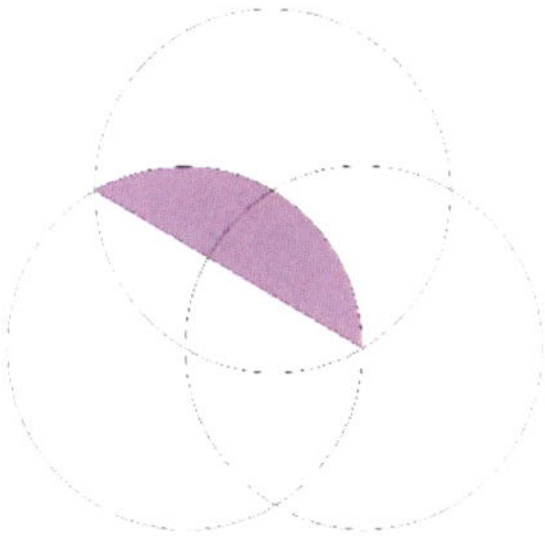

Figure B3
The Major SEGMENT6

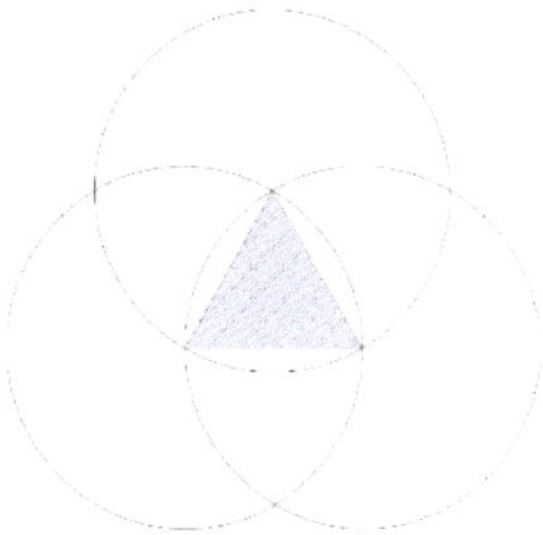

Figure B4
The Equilateral Triangle

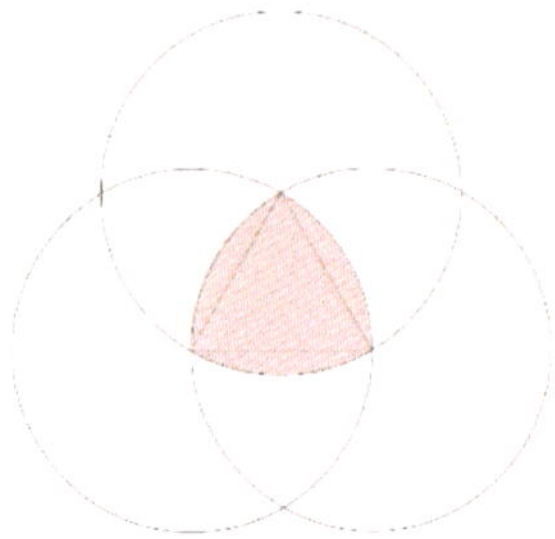

Figure B5
The Reuleaux Triangle (RT)

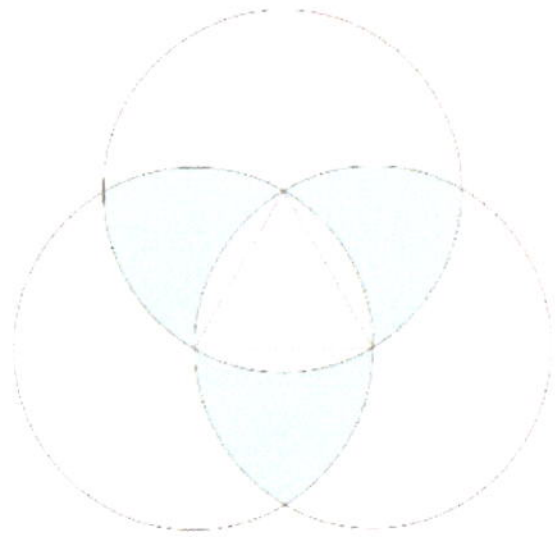

Figure B6
Linguoids

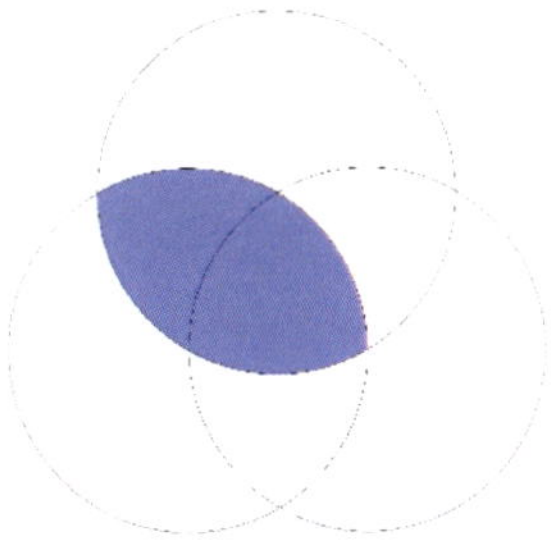

Figure B7
LENS3

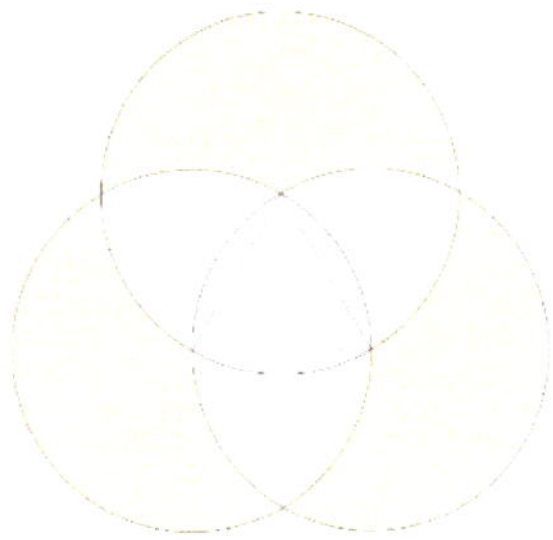

Figure B8
Lobes

APPENDIX C
SELECT LIST OF IMAGES

Figure:	1.3
Title:	The Anatomy of a Trefoil
Device:	Dell XLS
Processing:	Photodraw 2000 2.0.0.1129
Date of Creation:	10:38 3 October 2025
Creator:	James R Warren
Source:	This PC/DATA (D:)/TREFOIL/NewStock
Filename:	TrefoilDoubleRadiiAnotNoGrats.jpg

Figure:	1.4
Title:	Component Area Equations
Device:	Dell XLS
Processing:	Photodraw 2000 2.0.0.1129
	PTC MathCad Prime 3.1
	PaintNET v3.36
Date of Creation:	13:04 7 October 2025
Creator:	James R Warren
Source:	This PC/DATA (D:)/TREFOIL/NewStock
Filename:	TrefoilPlainHasEquationsForRisUnity.jpg

Figure:	1.5
Title:	GIMP Area Estimation Screen
Device:	Dell XLS
Processing:	Photodraw 2000 2.0.0.1129
	GIMP 3.0.2-1
	PaintNET v3.36
Date of Creation:	10:29 23 October 2025
Creator:	James R Warren
Source:	This PC/DATA (D:)/TREFOIL/PIX
Filename:	TREFOILreuleauxAreaSquarePixelEstimationScreenScreenshotANOT 2025-10-23 105700.jpg

Figure:	4.2
Title:	The Holy Trinity in a Trefoil
Device:	Dell XLS
Processing:	Photodraw 2000 2.0.0.1129
	GIMP 3.0.2-1
	PaintNET v3.36
Date of Creation:	11:07 1 November 2025
Creator:	James R Warren
Source:	This PC/DATA (D:)/TREFOIL/GRADEART
Filename:	TrefoilBlobOfThreeCirclesPlainRGBtrinityTransSmall.jpg

Figure:	5.1
Title:	The Reflection, Refraction and Transmission of A Ray of Light
Device:	Dell XLS
Processing:	Photodraw 2000 2.0.0.1129
Date of Creation:	11:02 1 November 2025
Creator:	James R Warren
Source:	This PC/DATA (D:)/TREFOIL/PIX/COLOR/ReflactionRefraction
Filename:	ReflectionRefraction.jpg

Figure:	5.3
Title:	The Subtraction (Absorption) of Reflected Colors
Device:	Dell XLS
Processing:	Photodraw 2000 2.0.0.1129
	PaintNET v3.36
Date of Creation:	14:56 25 October 2025
Creator:	James R Warren
Source:	This PC/DATA (D:)/TREFOIL/PIX/COLOR/AdditionAndSubtraction
Filename:	TrefoilBinariesRBGSubtractiveSystemColors.jpg

Figure:	6.1
Title:	The Mortuary Effigy of Hathor
	as Guide if the Dead
Device:	Unknown plate camera
Processing:	Unknown - halide plate
Date of Creation:	1923
Creator:	Harry Burton
Source:	Wikimedia Commons - 19:10 20 May 2022
Filename A:	11049.jpg.webp
Filename B:	Illustration from The Tomb of Tut-Ankh-Amen, Plate XVIII, Interior of Antechamber, the Hathor Couch.jpg
Figure:	6.2
Title:	The Head of the Bovine Hathor
	As Mehet-Weret
	From the Tomb of Tutankamun
Device:	Not Known
Processing:	Not Known
Date of Creation:	Not Known
Creator:	Not Known
Source:	https://egypt-museum.com/funerary-bed-of-tutankhamun-with-mehet-weret/
Filename:	Head-of-a-Funerary-Couch-in-th-form-of-Mehat-Werat.webp

Figure:	6.3
Title:	Author's Approximation of the Trilobates of Hathor
Device:	Dell XLS
Processing:	Photodraw 2000 Vers. 2.0.0.1129
Date of Creation:	10:35 7 November 2025
Creator:	James R Warren
Source:	This PC/DATA (D:)/TREFOIL/ART
Filename:	HathorTrilobateApproximatedOutline.jpg

Figure:	6.4
Title:	An Approximate Re-Constuction if the Hathor Trilobate
Device:	Dell XLS
Processing:	Photodraw 2000 Vers. 2.0.0.1129 PaintNET v3.36
Date of Creation:	10:35 7 November 2025
Creator:	James R Warren
Source:	This PC/DATA (D:)/TREFOIL/ART
Filename:	HathorTrilobateApproximatedInfilled.jpg

Figure:	6.9
Title:	A Lintel Supported by Columns
Device:	Dell XLS
Processing:	JuggernautXL 5.0
Date of Creation:	15:57 10 November 2025
Creator:	James R Warren (via text prompt)
Source:	https://dezgo.com/app/text2image_sdxl
Filename:	architraves_supported_by_Ionic_columns_around_a_Classical_Greek_temple_1776817289.png

Figure:	6.10
Title:	A Semicircular Arch
Device:	Dell XLS
Processing:	JuggernautXL 5.0
Date of Creation:	15:43 10 November 2025
Creator:	James R Warren (via text prompt)
Source:	https://dezgo.com/app/text2image_sdxl
Filename:	draw_a_semicircular_arch_with_a_keystone_ below_an_ashlar_bridge_1864326484.png

Figure:	6.11
Title:	A Cuspate Arch beneath a Bridge
Device:	Dell XLS
Processing:	JuggernautXL 5.0
Date of Creation:	15:57 10 November 2025
Creator:	James R Warren (via text prompt)
Source:	https://dezgo.com/app/text2image_sdxl
Filename:	draw_a_semicircular_arch_below_an_ashlar_bridge_1993625153.png

Figure:	6.12
Title:	The Parabolic Arch of Taq Kasra at Ctesiphon
Device:	Not Known - Victorian Plate Camera
Processing:	Not Known - Silver Halide
Date of Creation:	1864
Creator:	Not Known
Source:	Wikipedia "Ctesiphon" "Wonders of the Past" vol 2
Filename:	Ctesiphon-ruin_1864.jpg

Figure:	6.13
Title:	The Dona Maria Pia Bridge spanning the Duro at Oporto
Device:	Sony Cybershot
Processing:	Not Known
Date of Creation:	11:39 20 Novemeber 2007
Creator:	Jose Goncalves
Source:	Wikipedia "Maria Pia Bridge" This PC/DATA (D:)/TREFOIL/PIX
Filename:	Ponte_Maria_Pia_-_Porto.jpeg

Figure:	6.17
Title:	A Unitary Liguoid Approximates a Stable Parabola
Device:	Dell XLS
Processing:	EXCEL - MicroSoft Office 2021 Photodraw 2000 Vers. 2.0.0.1129 GIMP 3.0.2-1
Date of Creation:	20:42 9 November 2025
Creator:	James R Warren
Source:	This PC/DATA (D:)/Trefoil/New Stock
Filename:	ParabolaExEXCELWithLinguoidCrop.gif

Figure:	6.20
Title:	A Three-Light Lancet Window
Device:	Dell XLS
Processing:	EXCEL - MicroSoft Office 2021
	Photodraw 2000 Vers. 2.0.0.1129
	GIMP 3.0.2-1
	PaintNET v3.36
Date of Creation:	13:53 10 Novemeber 2025
Creator:	James R Warren
Source:	This PC/DATA (D:)/Trefoil/New Stock
Filename:	LancetWindow.jpg

Figure:	6.21
Title:	The Lady Chapel of Lichfield Cathedral
	From the East
Device:	Casio ZX1200
Processing:	None
Date of Creation:	09:10 8 November 2025
Creator:	James R Warren
Source:	This PC/DATA (D:)/TREFOIL/PIX
Filename:	Fig6dot20CIMG5561.JPG

Figure:	6.22
Title:	Trefoils and Rouleaux Triangles along Lichfield Cathedral's South Wall
Device:	Casio ZX1200
Processing:	None
Date of Creation:	09:27 8 November 2025
Creator:	James R Warren
Source:	This PC/DATA (D:)/TREFOIL/PIX
Filename:	Fig6dot21CIMG5568.JPG

Figure:	6.23
Title:	Inside the Lady Chapel
Device:	CASIO QV-5700
Processing:	None
Date of Creation:	10:29 3 March 2006
Creator:	James R Warren
Source:	This PC/DATA (D:)/TREFOIL/PIX
Filename:	LichfieldCathedralLadyChapel3Mar2006LADY1A.jpg

Figure: 6.24
Title: Trefoils on the North Wall of Lichfield Cathedral
Device: Casio ZX1200
Processing: GIMP 3.0.2-1
Date of Creation: 09:27 8 November 2025
Creator: James R Warren
Source: This PC/DATA (D:)/TREFOIL/PIX
Filename: Fig6dot24CIMG5614undist.jpg

Figure: 7.1
Title: The Association of Physical Parameters
In Mechanics
Device: Dell XLS
Processing: Photodraw 2000 2.0.0.1129
Date of Creation: 09:01 13 November 2025
Creator: James R Warren
Source: This PC/DATA (D:)/TREFOIL/MBTscience/Physics/System M-L-T
Filename: TrefoilMLTdynamicskinematicsAllNames.jpg

Figure:	7.7
Title:	The Sedimentary Trefoil
Device:	Dell XLS
Processing:	Photodraw 2000 2.0.0.1129
	PaintNET v3.36
	GIMP 3.0.2-1
Date of Creation:	09:01 13 November 2025
Creator:	James R Warren
Source:	This PC/DATA (D:)/TREFOIL/PIXtinted
Filename:	TrefoilBlobOfThreeCirclesPlainRYBAreniteArgilliteCarbonateTranss50.jpg

Figure:	7.9
Title:	Schiehallion Mountain from Loch Rannoch
Device:	CASIO QV-5700
Processing:	Photodraw 2000 2.0.0.1129
Date of Creation:	17:04 23 July 2006
Creator:	James R Warren
Source:	This PC/DATA (D:)/TREFOIL/PIX
Filename:	SCHI6sharp7small.jpg

INDEX